BUILDING INFORMATION MODELING FOR BEGINNERS
A SIMPLE GUIDE TO DIGITAL CONSTRUCTION

Goodluck Eromonsele

Title: **BUILDING INFORMATION MODELING FOR BEGINNERS**
A SIMPLE GUIDE TO DIGITAL CONSTRUCTION

ISBN: 979-8-89966-426-7

Author: Goodluck Eromonsele

Cover image: AI Generated

Publisher: Generis Publishing
Online orders: www.generis-publishing.com
Contact email: info@generis-publishing.com

Generis
PUBLISHING

BUILDING INFORMATION MODELING FOR BEGINNERS

A SIMPLE GUIDE TO DIGITAL CONSTRUCTION

GOODLUCK EROMONSELE

CONTENTS

Preface

Digital technology breakthroughs are driving a major revolution in the construction industry. Of these, the most revolutionary is Building Information Modeling (BIM), which is changing the way projects are designed, constructed, and managed. BIM is not just a tool or a software package; it is a methodology that fosters collaboration, enhances accuracy, and leads to more efficient and sustainable construction practices.

"Building Information Modeling for Beginners: A Simple Guide to Digital Construction" is written with the novice in mind. Whether you are a student, a young professional, or someone transitioning into the world of construction, this book aims to provide you with a solid foundation in BIM. The goal is to demystify the concepts and processes associated with BIM, making them accessible and understandable.

Throughout this book, I have endeavored to balance technical detail with practical application. The chapters cover everything from the basics of 3D modeling to the complexities of 7D BIM, which integrates sustainability into the construction process. You will also find discussions on the latest industry standards, as well as real-world case studies that demonstrate the impact of BIM on actual projects.

Writing this book has been an opportunity to reflect on the significant changes that BIM has brought to the construction industry. It is my hope that this guide will serve as a valuable resource for those beginning their journey with BIM, helping them to navigate the challenges and seize the opportunities that this innovative approach offers.

As you explore the pages of this book, I employ and encourage you to think about how BIM can be applied to your own work or construction project and how it might influence or enhance the future of the industry. BIM is more than just a trend; it is a fundamental shift or change in how we approach construction, and those who embrace it will be well-positioned to lead in the years to come.

Thank you for choosing this book as your guide to Building Information Modeling. I wish you great success and achievement in your learning journey and in all your future projects.

Goodluck Eromonsele
October 2025

Summary

"Building Information Modeling for Beginners: A Simple Guide to Digital Construction" is an essential resource for anyone seeking to understand and implement Building Information Modeling (BIM) in the construction industry. This book offers a detailed introduction to BIM, covering its foundational concepts, practical applications, and the significant impact it has on the construction process.

The guide is designed specifically for beginners, making complex ideas accessible and actionable. It explores the various dimensions of BIM, from 3D modeling to more advanced aspects like 4D (time management), 5D (cost estimation), and beyond. The book emphasizes the importance of integrating BIM into construction projects, highlighting how it can streamline workflows, improve collaboration, and enhance project outcomes.

Readers (students, researchers, and construction professionals) will find vast knowledge on how to adopt BIM in their own projects, including detailed explanations, best practices, and concepts that illustrate real-world applications. The book also delves into the benefits, application, and integration of related technologies associated with BIM adoption, offering strategies for quick understanding and steps in adopting BIM Workflow, process, and technology.

In addition to BIM, the book covers related digital tools and technologies that are transforming the construction industry, such as project management software, mobile applications, and digital surveying tools. It also addresses important topics like safety and risk management, sustainability, and the evolving standards and regulations that govern BIM practices.

By the end of this book guide, readers will have a solid understanding or comprehension of BIM and its role in modern construction. They will be equipped with the knowledge and tools needed to implement BIM successfully, whether they are working on small residential building projects or large-scale commercial or developments infrastructure projects. This book is an invaluable asset or resource for construction professionals, architects, engineers, local authorities, owners, and anyone interested in the future of digital construction.

Chapter 1: Introduction to Building Information Modeling (BIM)

Building Information Modeling (BIM) has become a crucial component of the construction industry that promotes efficiency, cooperation, and innovation. This introduction will give readers a strong foundation for understanding the BIM's concepts and principles, benefits and advantages, and applications. Through the use of BIM, professionals can improve project outcomes, streamline workflows, and adjust to the construction industry's changing environment or landscape.

In the dynamic and constantly evolving fields of design, engineering, and construction, Building Information Modeling (BIM) has emerged as a game-changing tool. BIM has changed how professionals approach designs, construction, and operation of buildings and infrastructure. This BIM overview aims to provide a complete understanding of all of its fundamental concepts and core ideas, applications, and benefits. Whether you're a professional trying to advance your skills and improve in your professional career or a student researching or exploring on this subject, this guidebook will give you a complete understanding of BIM and how it impacts the construction industry.

1.1 Revolutionizing the Construction Industry

The revolutionary technique known as "Building Information Modeling" (BIM) has completely transformed the architecture, engineering, construction, and operation (AECO) industry. It is a digital representation of a building's physical and functional characteristics, serving as a comprehensive and integrated platform for project stakeholders to collaborate and make quality informed decisions throughout the entire lifecycle of a project or a structure.

Beyond conventional 2D drawings, BIM includes a variety of information, such as geometry, spatial representation and relationships, computations and quantities, attributes and features, and performance data. BIM enables professionals to view, analyze, and simulate numerous aspects of the project by creating a virtual model of the building, enhancing design accuracy, improving construction efficiency, and streamlining facility management in building or structure operation phase.

The collaborative aspect of BIM is one of its most fundamental principles. It facilitates more effective communication and information or data exchange among architects, engineers, construction managers, manufacturers, contractors, and other stakeholders that promote interdisciplinary collaboration with the aid of the Industry

Foundation Classes (IFC) open file formats and guided by ISO 19650. As a result of this collaborative environment which ensures that all project stakeholders and concern parties work together seamlessly, reducing errors or fewer mistakes, conflicts or disputes, and project delays.

The benefits and advantages of BIM extend throughout the whole of a project lifecycle. BIM enables architects and engineers to create and evaluate multiple design alternatives, enhance, and optimize energy performance, and conduct simulations for lighting, acoustics, and thermal analyses during the project design phase. Construction managers, Builders and contractors can use BIM for construction sequencing, accurate quantity takeoffs and clash detection, leading to an improved coordination and cost control in construction projects. BIM provides an enormous quantity of information for facility management's operations, maintenance, repairs, and renovations, leading to more effective asset management and lower lifecycle costs.

With several governments and organizations (such as the U.K. government) regulating the use of BIM in public projects, the adoption of BIM has become more widespread in the AECO industry. Professionals can now produce sophisticated and intelligent 3D models, extract data for cost estimation, automate construction scheduling, and even incorporate sustainability studies and analysis as a result of the development of BIM software and tools.

In order to further improve project visualization, stakeholder engagement, and remote collaboration, emerging technologies like virtual reality, augmented reality, and cloud-based collaboration are being integrated into the BIM workflow. This trend has transformed the traditional construction industry into a digital one that optimizes the production of buildings and infrastructure in a sustainable manner.

1.2 Building Information Modelling (BIM) Definition

The three terms "Building," "Information," and "Modeling" make up the acronym BIM. The term "building" presents a problem from the standpoint of design, as the field needs a construction protagonist to strengthen the allegedly deteriorating relationship between the architect and the contractor and the general building organization (Briscoe 2016). While "Information" refers to parametric design data and other information related to architecture, engineering, and construction, "Modeling" refers to the process of transforming ideas into forms or shapes that are visible in spatial representation and beyond.

Building information modeling (BIM) is a digital technique for creating, managing, and disseminating accurate comprehensive information on a building or infrastructure project throughout its whole lifecycle and existence. Design, construction, and operation of a project are all integrated into a single collaborative platform using BIM.

Antonopoulou and Bryan 2017, (p. 1) states that; "*The term Building Information Modelling (BIM) is used to describe a collaborative process for the production and management of structured electronic information. BIM is not new technology, its origins being in object-based parametric modelling applications for mechanical systems design in the 1980s.*" While Eastman et al., 2008 cited in Ingibjorg et al. 2017 (p.15) viewed BIM as "*A modelling technology and associated set of processes to produce, communicate and analyse building models.*"

BIM is defined by BuildingSMART International, which was cited by Ingibjorg et al. in 2017 (p.15) as; "*A new approach to being able to describe and display the information required for the design, construction and operation of constructed facilities.*"

BIM's fundamental component is a 3D model-based system with intelligent objects or components. These components contain information regarding their physical and functional characteristics, which includes dimensions, materials, maintenance requirements, and performance information. Additional data can also be integrated using BIM, including sustainability analyses, schedules, and cost estimates or cost predictions.

Enhancing communication, coordination, and collaboration among all project stakeholders is the primary objective of BIM. Working collaboratively on a common digital model enables Architects, Engineers, Contractors, and Facility managers to visualize the project, detect and resolve any potential disputes or issues, and make well-informed decisions. Project teams can increase productivity, minimize mistakes and rework, optimize resource allocation, and improve overall project outcomes by using BIM. Better visualization, simulation, and analysis are made possible by BIM, which improves facility management, construction sequencing, and design accuracy. Additionally, it supports the implementation of sustainable and energy-saving techniques and practices.

The way projects are planned, executed, and maintained has been transformed by BIM, which has become a crucial component of the construction industry. It makes it easier to develop and operate buildings holistically, enhancing outcomes for projects and reducing costs throughout a facility's full lifecycle.

1.2.1 BIM Maturity Levels and Dimensions definition

Building Information Modeling (BIM) maturity levels and dimensions refer to the degrees and phases of development and components of BIM application in the construction industry. These levels and dimensions aid in assessing how organizations are utilizing and adopting BIM technologies and processes.

The gradual and continual development of quality, repeatability, and predictability within available boundaries of BIM capability is known as BIM Maturity (Succar 2016). And on the other hand; Aung (2004) states that, the type and details of the digital building information required in a BIM model for a construction project are referred to as "BIM dimensions."

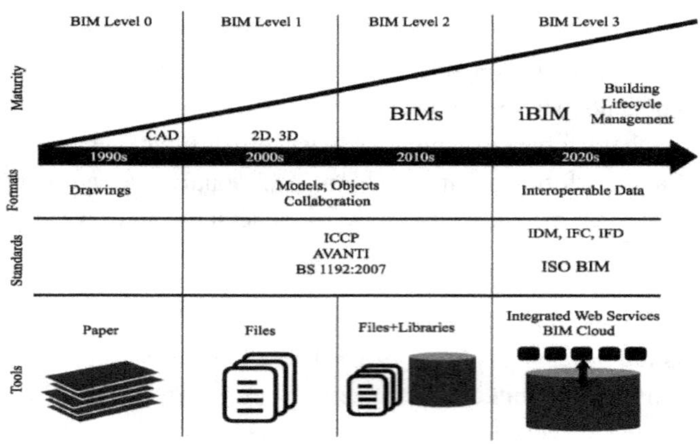

Figure 1: BIM Level 1-3 (Source: AlMashjary, Zolkafli, and Razak 2020)

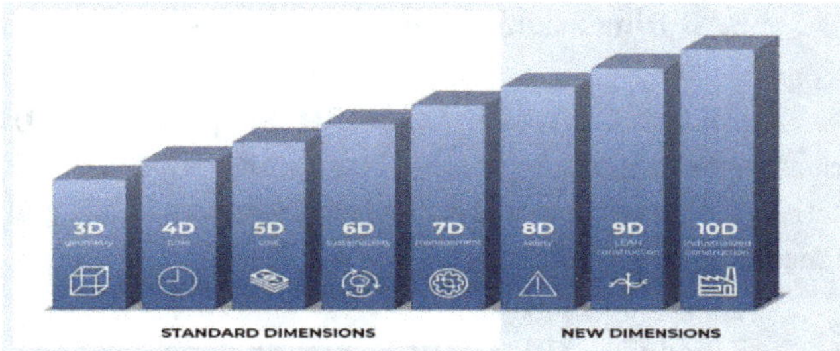

Figure 2: The BIM Dimensions (Source: Melovics 2022)

A. **BIM Maturity Levels**

- **Level 0**: Traditional 2D drawings and documents are used for project documentation and communication at this level instead of BIM. ICT applications are fundamentally used to create drawings and drawn components utilizing CAD tools (AlMashjary, Zolkafli, and Razak 2020).

- **Level 1**: The introduction of basic 3D modeling results in little collaboration and fragmented information transmission and data exchange method. The project team's usage of the Common Data Environment (CDE) to manage and share information implies partial collaboration.

- **Level 2**: At this stage, an environment of collaboration where several disciplines can work together. Each discipline develops and manages a separate set of 3D models, which are subsequently integrated for coordination. In order to ensure collaboration between all disciplines and the creation of an integrated and interoperable data collection, this highly sophisticated 3D modeling technique with connected data takes use of a shared Common data environment (CDE) and standards.

- **Level 3**: At this level, integrating and collaborating are the main focuses. One shared 3D model in a Common data environment (CDE) is utilized by all project stakeholders and participants, enabling real-time collaboration and information exchange and sharing.

B. BIM Dimensions: Standard and New Dimensions

- **Spatial Coordination or 3D Shape (BIM-3D)**: Buildings or infrastructure projects are represented by virtual 3D models of their physical characteristics, geometry, spatial representation, relationship, and linkages in the three dimensions of BIM.

- **Scheduling (BIM-4D)**: The fourth dimension, which incorporates time into the three-dimensional model, makes it possible to visualize and simulate project progress, development, and construction sequencing. It aids in planning, scheduling, and managing of construction projects.

- **Estimating (BIM-5D)**: The fifth dimension (5D) incorporates cost-related data into the third-dimensional (3D) model, enabling the project stakeholders to carry out cost estimation, quantity takeoffs, and cost tracking throughout the project lifecycle.

- **Sustainability (BIM-6D)**: The sixth dimension includes the BIM model's incorporation of sustainability and environmental data. It supports sustainable design and decision-making by facilitating energy analysis, carbon footprint assessment, and lifecycle analysis.

- **Facility Management (BIM-7D):** Facility management and operation are the focus of the seventh dimension. To support effective facility management, it also incorporates asset management information, maintenance schedules, and equipment specifications into the BIM model.

- **Safety (BIM-8D):** BIM-8D, also referred to as Building Information Modeling for Safety, is a technique that facilitates the use of BIM technology's capabilities to enhance safety in the construction industry. A comprehensive safety management system is produced by BIM-8D, which expands on the traditional 3D modeling aspects of BIM by incorporating additional safety-related dimensions such as time (4D), cost (5D), and sustainability (6D). Construction projects can benefit from improved safety planning, hazard identification, clash resolution, and enhanced communication among project stakeholders by incorporating safety considerations into the BIM process through BIM-8D. This eventually improves the safety at workplace and reduces the probability and likelihood of accidents and injuries in the construction industry.

- **Lean Construction (BIM-9D):** Lean construction is an ideology that emphasizes maximizing value, reducing waste, and enhancing collaboration and efficiency throughout the construction process in order to optimize project delivery. BIM technology can, however, be integrated into lean construction practices to enhance their effectiveness through visualization and communication, value stream mapping, clash detection and coordination, prefabrication and modularization, lean scheduling streamlining workflows, data-driven decision making, and continuous improvement.

- **Industrialized Construction (BIM-10D):** The prefabrication of building components in regulated or controlled factory and manufacturing environments and their subsequent assembly on-site constitute industrialized construction, also known as off-site or modular construction. The use of Building Information Modeling (BIM) technology can greatly support and improve Industrialized Construction practices. Although the term "BIM-10D" is not widely utilized or recognized. Design and visualization, clash detection and resolution, component standardization and optimization, prefabrication planning, supply chain management, and assembly simulation and sequencing are all capabilities that BIM can offer to industrialize construction.

These maturity levels and dimensions create a framework for comprehending BIM's progressive adoption, implementation, and use across the lifecycle of construction projects, facilitating better and improved collaboration, coordination, and decision-making.

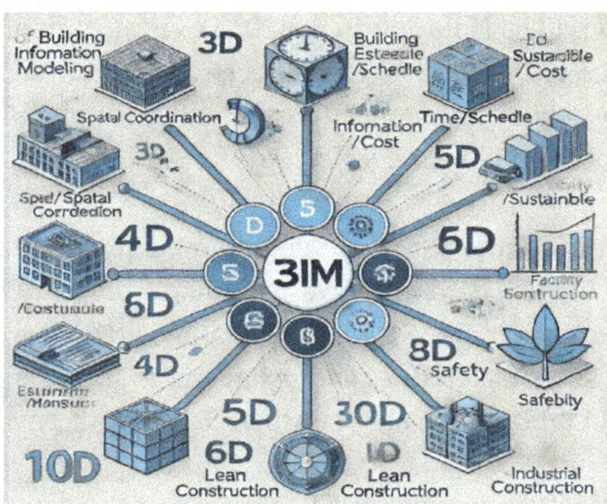

Figure 3: BIM dimensions and functions

1.3 Recognizing and understanding the fundamentals of Building Information Modelling

Understanding the fundamentals of Building Information Modeling (BIM) means becoming familiar with its key components and guiding principles. The Design, Construction, and Operation (DCO) sector's extensive understanding domain has become known as "Building Information Modelling" (Succar 2010). It's crucial to comprehend the basics of BIM as a recent graduate in the fields of architecture, Engineering, construction, and operation or as a professional who hasn't used BIM. Below are the core components or fundamental aspects of BIM:

1) **Digital Representation**: The primary purpose of Building Information Modelling is to create a digital representation of a particular typical building or infrastructure project. It generates a virtual model of the project's physical and functional characteristics aspects using 3D modeling software.

2) **Information Integration**: Various types of information are integrated into the digital model through BIM. This consists of both geometric and non-geometric data, such as material properties, cost estimates and projections, schedules, and performance specifications. Geometric data include dimensions and spatial relationships.

3) **Collaboration and Coordination**: Project stakeholders' collaboration and collaboration are emphasized and prioritized by BIM. The information from various systems, such as the mechanical, plumbing, and electrical systems, are able to be analyzed using BIM to determine if there could be any potential clashes or issues (The Hourigan Team 2013). On a common platform, specialists from numerous disciplines, like architects, engineers, contractors, and facility managers, may collaborate to share information, spot conflicts or clashes, and solve issues more quickly and efficiently.

4) **Parametric Objects:** BIM contain intelligent digital parametric objects. In the digital model, BIM employs parametric objects or components. These objects are intelligent, and they can communicate and store information about their properties, characteristics, and behaviors. For example, a door or window object can store information about its material, size, thermal properties, and cost.

5) **Lifecycle Perspective**: BIM takes into consideration every stage of a building's lifecycle, from design and construction through operation and

maintenance. Throughout the project's lifecycle, the digital model serves as a central repository of data that can be accessed and updated, facilitating decision-making and minimizing inefficiencies in a typical construction project.

6) **Visualization and Simulation:** Realistic simulation and visualization of a project are made possible by BIM. Owners are able to visualize the project and, as a result, offer more profound feedback on what they want compared to the traditional 2D drawings that have been utilized in the past (The Hourigan Team 2013). Stakeholders could more effectively assess design options, conduct simulation performance analysis and assessments, and visualize the building in a virtual environment setting as a result of the 3D model.

7) **Data Interoperability**: The ability of two or more systems to interact and exchange information with one another is known as data interoperability (Santos 2010). BIM encourages data interoperability, enabling seamless information exchange between different software applications and programs. This prevents data loss and duplication and ensures that data generated in one phase of the project can be readily and easily used in subsequent phases.

8) **Standardization:** Industry standards are relied on by BIM to guarantee consistency and interoperability. Data formats and exchange protocols for BIM models are defined by standards like the Industry Foundation Classes (IFC) and COBie (Construction Operations Building Information Exchange).

Construction professionals can effectively use BIM to enhance collaboration, improve project outcomes, play active stakeholder roles, and optimize the lifecycle management of buildings and infrastructure projects by understanding and comprehension of these fundamentals.

1.4 The history and evolution of BIM

Building Information Modelling history and evolution can be traced back spanning to several decades. Many authors have explored at the evolution and history of BIM.

BIM's origins can be traced back to 1975, when American professor Charles M. Eastman published a description of a working prototype in the AIA Journal, according to the Alison 2023 BIM Course Module1: Introduction to Building Information Modeling (BIM). Describing interaction definition pieces that incorporate data from sections, facades, viewpoints, and maps. In the late 1970s and early 1980s, attempts were made to use BIM commercially in Finland and Great Britain, where workstation products like GLIDE, RUCAPS, Sonata, Reflex, and Gable 4D Series were among the first software tools developed for modeling buildings. Alison BIM module course further states that studies similar to this were also conducted in Europe in the 1970s and 1980s; After acquiring Revit Technology Corporation in 2002, and in 2003 Revit replaced the QWG format, which had been the cornerstone for Autodesk's subsequent developments for 20 years. Below is an overview and synopsis of its key milestones as they were examined in various literary works:

	Overview	Milestones
1	Early stages of development (1960s-1980s):	Early initiatives in computer-aided design (CAD) and computer-aided engineering (CAE) in the 1960s laid the groundwork for BIM. The foundation for digital modeling and simulation was laid by these technologies. Researchers and pioneers explored the ideas of parametric modeling and object-oriented programming, which would later become crucial elements of BIM in the 1970s and 1980s.
2.	An Introduction of the Term "Building Information Modelling" (1990s):	In the 1990s, the phrase "Building Information Modelling" was coined to characterize or describe the then-emerging approach to utilizing digital models to collect and manage building information. During this time, the emphasis switched from 2D CAD drawings to 3D models that were data-rich.
3	Adoption by Industry (2000s):	BIM gained momentum in the early 2000s as a result of improved software capabilities, increasing computing power, and growing public awareness of the benefits it provides. Governmental programs and initiatives in countries like the United States of America and the United Kingdom played a major role or factor in promoting the adoption of BIM in public projects. Building owners, architects, engineers, construction managers, and contractors began to recognize its potential to enhancing project outcomes, improve collaboration, and reduce errors.
4	BIM capabilities expansion (2010s):	BIM capabilities saw considerable developments in the 2010s. The introduction and invention of cloud computing and mobile technology made it possible to collaborate in real-time and access BIM models from any location. With the addition of capabilities like clash detection, energy analysis, and construction sequencing, BIM software became more sophisticated. In addition, BIM also encompasses facility management and operation extending beyond design and construction.

5	Interoperability and Open BIM (2010s):	As BIM developed, interoperability became a crucial key focus in the evolution. The introduction and development of open standards like BuildingSMART and the Industry Foundation Classes (IFC) encouraged and promoted data exchange and interoperability between various software systems or platforms. This made it possible for project stakeholders to collaborate seamlessly and facilitate the incorporation or integration of BIM into different project phases.
6	Level of Development Framework for BIM (2010s):	In order to define and standardize the level of detail and reliability of information in BIM models at different project stages, the BIM level of Development (LOD) framework was developed and introduced. This framework ensures consistent communication and expectations among project participants by defining or providing a common understanding of what may be expected from a BIM model at a specific LOD.
7	BIM Future and upcoming trends:	Innovative advances in technology like augmented reality (AR), virtual reality (VR), and artificial intelligence (AI) are likely to further involve in BIM advancements in the future. These technologies have the prospect to enhance visualization, simulation, and automation within the BIM process and also have the potential to improve project outcomes and decision-making. Additionally, BIM has the potential to integrate with tools for life cycle assessment (LCA) and supports the advancement of digital twin technologies for buildings currently.

Table 1: BIM Historical overview with Milestones

The development and advancement of BIM from simple computer-aided design to a comprehensive collaborative process, that spans the entire lifecycle of a building or infrastructure project is showcased as a whole in its history. The construction industry has undergone a revolution as a result of BIM, which has promoted or enhanced efficiency and productivity, reduced costs, and improved project results.

1.5 Distinguishing between traditional CAD and BIM approaches

The scope and functionality of traditional or conventional Computer-Aided Design (CAD) and Building Information Modeling (BIM) process approach and techniques are different. The use of computer technology for designing physical real and virtual objects, such as 2D and 3D architectural models, is known as CAD.;

whereas BIM is a human activity that entails significant process modifications in construction, it is not a sort of software (Hussien 2023). The key differences between these two methods or approaches are listed below:

	Features	Traditional CAD	BIM approaches
1	Representation	Primarily CAD concentrates on 2D drawings or 3D models that depict the geometry of building elements. It draws emphasis on the project's visual representation	There is information associated to architectural or building elements, such as their properties, relationships, and behavior throughout the project's lifecycle, is additionally included in BIM, in contrast to the CAD, which encompasses just architectural geometry.
2	Information Integration	Beyond the graphical appearance and representation, CAD models generally lack comprehensive or substantial information.	However, BIM integrates multiple kinds of information, like operational, functional, and physical data, into the model. Materials specifications, performance characteristics, cost estimates, schedules, and other details may be included in this information.
3	Collaboration and Coordination	The CAD systems are frequently used singularly by specific professional disciplines, resulting in isolated workflows.	On the contrary, BIM promotes collaboration and coordination among all project stakeholders. It creates a shared platform or common working environment for multiple disciplines to work together, facilitating improved communication, clash detection, and conflict identification and resolution.
4	Lifecycle Perspective	The primary applications of CAD are for design and drafting purposes. It focuses on creating intricate and	BIM, on the other hand, takes a comprehensive lifecycle approach. It includes every stage of a

		detailed drawings or models at specific or particular project stages.	project's lifecycle, including from design and construction to operation and maintenance. Throughout all phases of the project, BIM models act as a central repository for information that can be accessed and updated.
5	Parametric Objects	CAD models mainly consist of geometric entities with little to no intelligence.	Parametric objects or components, which have both geometric and data attributes, are employed in BIM models. These objects keep records of their properties, behaviors, and relationships with other elements. When one object is modified, all related elements of the model are automatically updated, ensuring accuracy and consistency.
6	Analysis and Simulation	CAD tools might offer restricted or limited capabilities for analysis.	BIM provides improved functionalities for analysis and simulation. Construction sequencing, cost estimation, energy analysis, clash detection, and other simulations can all be performed using BIM models. As a result, stakeholders could evaluate design alternatives, detect conflicts, and optimize performance.
7	Interoperability	Data exchange across different software platforms can be difficult or challenging since CAD systems often have poor or limited interoperability.	BIM adheres to open standards such as the Industry Foundation Classes (IFC) emphasizing data interoperability or a strong emphasis on data interchange. This makes it possible for easy and seamless data exchange

			across different BIM software programs and applications, fostering collaboration among project participants.

Table 2: Difference between BIM and CAD

Overall, BIM encompasses a more comprehensive strategic approach that integrates information, collaboration, and lifecycle management, while CAD focuses largely on graphical representation. BIM's emphasis on data-rich models and collaboration leads to improve and better project coordination, efficiency, and decision-making throughout the project lifecycle.

Chapter 2: Core Concepts and principles of Building Information Modeling

Building Information Modeling (BIM), which was previously covered in the introduction chapter, is a process and software, not just a particular piece of software. Technology, Process, and Policy are the three main fields of BIM (Succar 2010). The four pillars of BIM, according to Future Learn 2023, which is affiliated with Coventry University, are policy, people, process, and technology. The policy, technology, people, and process pillars of BIM will be used to illustrate and discuss the core concept of BIM in this section. Some core concepts of BIM are 3D Modeling, Information Integration, Parametric Design, Collaboration and coordination, Data Analysis and Simulation, Lifecycle Management, Visualization and communication, and Standardization and Interoperability.

Three-dimensional (3D) modeling is one of the fundamental concepts in BIM. A virtual representation or depiction of a building or infrastructure project is created by BIM using 3D modeling software. The project's components, such as walls, floors, doors, windows, and structural elements, are detailed and described in detail in this 3D model.

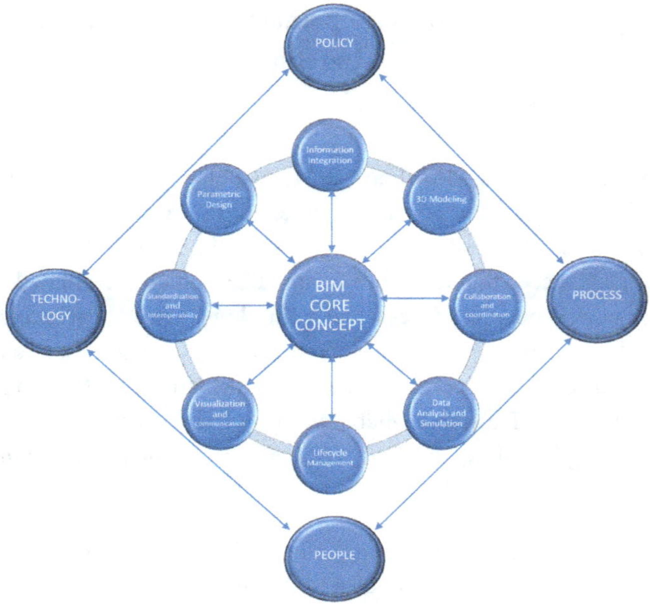

Figure 4: BIM core concept and pillars

Information Integration is yet another core BIM concept. Multiple forms or types of information are integrated into the digital model through BIM. In addition to geometric data, it also incorporates non-graphical data like costs, material specifications, performance characteristics and traits, and scheduling information. Through this integration, decision-making is enhanced throughout the project lifecycle.

Another concept is parametric design. BIM makes use of parametric design principles or concepts, which means that model elements or components are interconnected and influenced by specific parameters or coupled to one another and affected by particular factors. For instance, if the size of a room is altered, all associated elements, like the walls, doors, and windows, will be adjusted automatically to ensure accuracy and consistency.

Yet another crucial BIM concept is collaboration and coordination. By providing project's stakeholders and everyone involved in a particular project access to a commonly shared platform for exchanging information, BIM promotes or fosters collaboration amongst project participants or project stakeholders. Reduces errors and conflicts by enabling real-time collaboration, coordination, and communication between architects, engineers, contractors, and other stakeholders or parties involved.

Data Analysis and Simulation is core fundamental in BIM. BIM enables data analysis and simulation simpler by facilitating a variety of purposes. To detect and resolve conflicts or optimize designs before construction commences, it can simulate energy performance, lighting analysis, structural analysis, and clash detection.

Lifecycle Management features remain an important concept of BIM. A construction project's whole lifecycle, from design and construction to operation and maintenance, is supported by BIM. It enables better asset management and maintenance planning by allowing the seamless transmission of information from one phase to another.

BIM also has a unique concept known as "Visualization and Communication." BIM offers visual representations of the project, allowing stakeholders to more effectively comprehend and communicate design intent. It facilitates client presentations and project communication by enabling realistic renderings, walkthroughs, and virtual reality experiences.

Lastly, Standardization and Interoperability Standard guideline is another BIM concept. BIM emphasizes and promotes standardization and interoperability between various software tools and disciplines or fields of study. To guarantee and ensure compatibility and seamless data exchange and transmission, it relies on open data standards like Industry Foundation Classes (IFC) and Construction Operations Building exchange (COBie).

These core BIM concepts strengthen and promote collaboration, improve project outcomes, reduces or cut costs, and boost overall efficiency and effectiveness in the construction industry.

2.1 BIM components: objects, parameters, and relationships

Objects, parameters, and relationships are the basic components or fundamental elements of digital models in the context of Building Information Modeling (BIM). In order to create detailed, accurate, comprehensive, and information-rich representations of buildings, these components or elements are essential. These are the basic constituent or parts of the BIM system. BIM potent is defined by the fusion or combination of these components. The breakdown of each component is as follows:

A) Objects: In BIM, the term "object" refers to individual elements or components that make up the building model. Various physical and functional aspects of a building, such as the walls, doors, windows, columns, HVAC systems, electrical fixtures, and more, are represented by these objects. Each object is defined with specific properties, traits or characteristics, and geometry which correspond or reflects to its actual or real-world counterpart. They can also take the form of a "generic" or "specific" object, provided in open exchange, platform-neutral file formats, such as IFC, or made available in a variety of file formats suited for use in software like Revit Architecture and Graphisoft ArchiCAD.

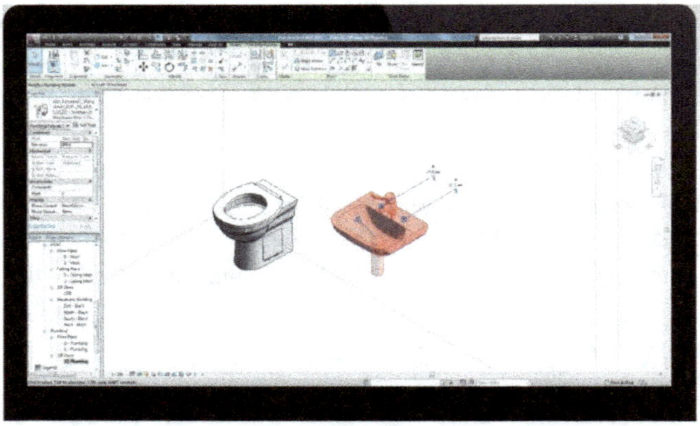

Figure 5: BIM objects in Architecture design (Source: McPartland 2017

B) Parameters: The characteristics or attributes that are assigned to objects in BIM are called parameters. Users can input and retrieve information related to the objects' properties and behaviors due to these definitions or specifications. In addition, parametric components include both geometric (such as length, width, and height) and non-geometric (e.g., material, manufacturer, cost, and performance ratings) properties. The data required or necessary to describe and analyze objects in the model is provided through parameters.

C) Relationships: The connections and dependencies between objects in a BIM model are described by relationships. These connections or relationships can be functional, spatial, or logical. In three-dimensional space, spatial relationships define how objects are placed or positioned and interact with each other. Functional relationships describe or capture how objects interact or affect one another or each other's behavior in a BIM model, such as the way or manner HVAC systems interact with rooms. Logical relationships create or establish links based on specific rules, regulations, or standards criteria, enabling automated processes and data consistency.

BIM models become dynamic and intelligent representations of buildings by leveraging these components. The capability to attach detailed information to each element is made possible by objects that are defined with specific parameters. The model is able to accurately depict or represent the interactions and dependencies between building components in reality or real-world because due to relationships between objects. Throughout the building lifecycle, collaboration, analysis, and decision-making are facilitated due to this information-rich and integrated model.

2.2 The importance of data consistency and accuracy in BIM

The quality, homogeneity, and dependability of information made feasible by BIM can be referred to as data consistency and accuracy in BIM. Consistency and accuracy of data are critical aspects in Building Information Modeling, which have a significant impact on the effectiveness and reliability of BIM processes and workflows. For this reason, it is crucial that the data provided by the BIM database be reliable or trustworthy and of high quality. Explanations of some reasons why data accuracy and consistency are crucial and important in BIM:

1) **Analysis and Decision-making**: BIM models are useful or serve as valuable sources of information or tools for conducting various analyses and coming to informed decisions. Data consistency ensures the model's data is reliable or trustworthy. And uniform or consistent across different elements or all of its components and disciplines. In order to execute accurate simulations, energy analysis, clash detection or collision detection, cost estimations or predictions, and other assessments, enabling stakeholders to make well-informed decisions, precise or reliable data is needed or required as a solid foundation.

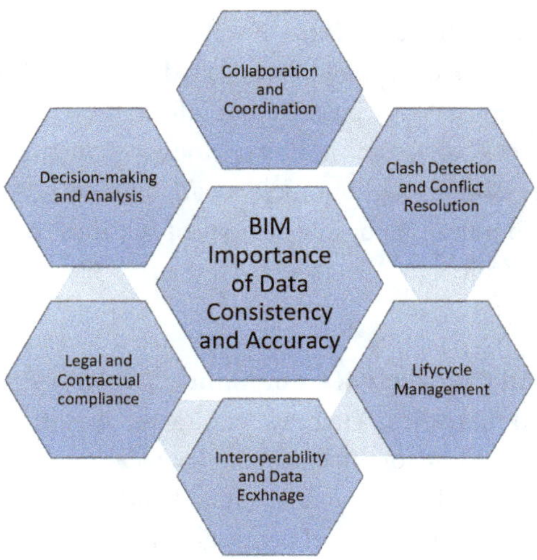

Figure 6: Importance of BIM data consistency

2) **Coordination and Collaboration**: BIM promotes collaboration among project participants, allowing them to work on a shared model. Data consistency ensures that all stakeholders are referring to the same and up-to-date information, reducing confusion and conflicts. Accurate data ensures that the model accurately represents the real-world conditions, supporting effective coordination among different disciplines and minimizing errors and rework.

3) **Conflict Resolution and Clash Detection**: BIM facilitates clash detection, which involves identifying conflicts or clashes between different building elements (e.g., pipes conflicting with structural members). Data consistency ensures that clashes are accurately detected, preventing clashes caused by discrepancies or inaccuracies in the model. Accurate data helps in resolving clashes effectively, enabling design modifications or alternative solutions to be implemented efficiently.

4) **Lifecycle Management**: BIM extends beyond the design and construction phases to encompass the entire lifecycle of a building. Data consistency and accuracy are crucial for managing the building throughout its operational phase. Reliable and accurate data ensures that facility managers have access to up-to-date and reliable information about the building's components, maintenance schedules, warranties, and other relevant data, facilitating effective facility management and maintenance activities.

5) **Interoperability and Data Exchange**: BIM relies on interoperability, allowing data to be exchanged between different software applications and platforms. Data consistency ensures that information is correctly transferred and interpreted across different software tools, maintaining the integrity and accuracy of the data throughout the exchange process. Accurate data facilitates seamless collaboration and integration of various systems and disciplines involved in the project.

6) **Legal and Contractual Compliance**: BIM models may be used for legal and contractual purposes, such as contracts, specifications, and regulatory compliance. Inaccurate or inconsistent data can lead to legal disputes, non-compliance with regulations, and financial liabilities. Ensuring data consistency and accuracy in BIM models helps mitigate risks and ensures compliance with contractual obligations and legal requirements.

By maintaining data consistency and accuracy in BIM, stakeholders can rely on the model as a trustworthy source of information throughout the building lifecycle. This enhances decision-making, collaboration, clash detection, and overall project efficiency, leading to improved project outcomes and reduced costs and risks.

2.3 Building a BIM model and the concept of federated models

The process of creating and integration or combination of various digital representations of a building's physical and functional elements and aspects together into a unified and single digital form or model is known as Building a BIM (Building Information Modeling) model. The use of federated models is the key concept in this BIM process or strategy. In BIM projects, federated models are essential for simplifying collaboration, collision and clash detection, coordination, and data integration. They contribute to enhancing the overall project efficiency, project outcomes, and more importantly, the design and construction process of a building project. Below is a description or explanation of the process of creating or building a BIM model and the concept or strategy of federated models below:

2.3.1 Creating or Building a BIM Model

1. Gathering Information: The first step in the process is gathering pertinent and relevant information regarding the building or construction project, such as architectural drawings and plans, structural designs, mechanical and electrical systems, and other pertinent data. The development of the BIM model is based on this information or serves as the basis for developing the BIM model.

2. Creating Discipline-Specific Models: Different disciplines develop their own discipline-specific models, such as architecture, structure, and MEP (mechanical, electrical, and plumbing). These models capture the necessary details and information and focus on the unique aspects of distinctive features of each discipline.

3. Coordinating and Collaborating: A comprehensive BIM model is then created by coordinating and integrating the discipline-specific models into one. Collaboration among project stakeholders is involved, such as the architects, engineers, construction manager, contractors, and subcontractors. To identify and resolve conflicts amongst the various building elements in the model, clash detection and resolution processes are carried out.

4. Data Integration: A common data environment is used to connect or link together the discipline-specific models, allowing or enabling data integration and exchange. By ensuring information flows smoothly or seamlessly between different parts of the model, keeping and maintaining data consistency and accuracy.

2.3.2 The Federated Models Concept

Federated models are BIM models that combine various or multiple discipline-specific models into a single all-encompassing and comprehensive BIM model. A particular aspect of the building or structure, such as the architectural design, structural analysis, or MEP systems, is represented by each discipline-specific model. These individual models are brought together in the federated model to produce or create an integrated and coordinated representation of the entire building or construction project. Federated models' key features or aspects include:

1. Collaboration: Federated models allow for collaboration amongst the various building project disciplines. Stakeholders can spot, detect and resolve clashes, coordinate their efforts, and make sure that all aspects or elements of the building are well-coordinated and aligned by integrating the various models.

2. Clash Detection and Resolution: Federated models make it easier to detect and resolve clashes. Conflicts or clashes between elements from various or different disciplines can be identified and addressed before construction start or begins by integrating the discipline-specific models. This minimizes or reduces errors, rework, and costly on-site clashes.

3. Integration and Coordination: Federated models make it possible to coordinate and integrate different design and engineering disciplines or professions. The models provide the project stakeholders a holistic or comprehensive view of the building or construction project, enabling them to comprehend how the different elements interact and impact with each other.

4. Data Consistency: Data consistency across different discipline-specific models is ensured by federated models' concept. Data can also be shared or exchanged and updated in real-time by integrating the models into a single or unified BIM model, ensuring and maintaining consistency throughout the design and construction process.

Chapter 3: BIM Software and Tools

Building information modeling (BIM) software and tools are a set of computer programs created as instructions to computers to help with the design and creation of building models and to assist construction or building projects in a BIM process. BIM is a methodology made viable by sophisticated computer tools (Luciani, Garagnani, and Mingucci 2012). Software and tools for BIM (Building Information Modeling) are designed to make it easier to create, manage, and share digital representations of building or infrastructure projects. Throughout the course of the project lifecycle, they make it possible for various stakeholders, including architects, engineers, construction managers, contractors, and clients, to collaborate together more effectively. Autodesk Revit, Navisworks, ArchiCAD, Solibri, Tekla Structures, BIM 360, Dynamo, Bluebeam Revu, BIMobject, Bentley AECOsim, Vectorworks Architect, Trimble SketchUp, Nemetschek Allplan, and Graphisoft BIMx are some examples of frequently used BIM software platforms and tools, with their explanations and overview provided below:

1. **Autodesk Revit**: Revit is a popularly and widely used BIM software platforms and tool that enables or allows users to design or create and manage 3D models of buildings or structures in construction projects. It provides or offers comprehensive tools for architectural design, MEP (mechanical, electrical, and plumbing) systems design, structural engineering design, and construction documentation.

   Revit is a well-known and established BIM platform, in accordance with Alison's BIM course 2023; Revit offers a user-friendly interface with drag-over indicators for each function and a smart cursor. Its operator menus grey out operations that are not available in the context of the current system and are sensibly organized according to workflow. Additionally, it offers two-way direction editing from schedules for doors, door hardware, and other items, as well as two-way direction editing from drawings to and from the model. Revit is a powerful BIM platform because of its variety of additional or extra applications.

2. **ArchiCAD**: Another well-liked BIM software that offers tools for creating or producing, and managing building models is ArchiCAD. It provides features or offers tools for collaboration, 3D modeling, architectural design, and construction documentation. A BIM platform called ArchiCAD is renowned for its user-friendly workflows and intuitive user interface. It provides tools for 3D

modeling, construction documentation, and architectural design. Early-stage design is emphasized or prioritized by ArchiCAD, which includes features like energy analysis and virtual building exploration.

According to Alison BIM course 2023; ArchiCAD is the oldest BIM tool platform for architectural design. Alison BIM course 2023 further expatiates that the user interface of ArchiCAD is well-designed and includes an intelligent cursor, drag-over operator hints, and context-sensitive operator menus. Its devoted user base enjoys its user-friendliness and model generation. Drawing generation is automatically managed by the system in ArchiCAD. Plans, 3D documents, sections, and elevations can all be edited or modified in both directions.

3. **Navisworks**: Software for BIM coordination and clash detection is called or known as Navisworks. It enables users or project stakeholders to combine models from different disciplines of various professionals, spot or identify clashes or conflicts, and more efficiently and effectively manage and coordinate project activities.

4. **Tekla Structures**: Tekla Corporation produced Tekla structures. The software was designed and programmed to enable collaboration among users or project stakeholders on a single server-based model of a building or a construction project. Primarily, Tekla Structures are designed for structural engineering and detailing works. It offers or provides a cutting-edge and advanced tool for modeling and analysis of steel, concrete, and timber or wooden structures. In addition, it allows and supports data integration and collaboration with other BIM software.

Tekla structures is a powerful tool that can be used to model structures that incorporates a variety of structural materials and detailing, its complete functionality is extremely challenging or difficult to comprehend or understand and use to its maximum potential or fullest extent (Alison BIM course module 2023). Tekla Structures facilitates the processes for detailing, fabrication, and construction workflows, enabling engineers, fabricators, and contractors to collaborate effectively.

5. **Solibri**: A BIM quality assurance software called Solibri assists with rule-based analysis, clash detection and identification, and model checks. It guarantees adherence to standards and guidelines, rules or regulations, and project-specific requirements or needs. A group of individuals who saw the world changing, particularly in building technologies, formed Solibri. And in

1999, a modeling program was created with the intention of enhancing and validating common standards in construction software and digital construction technologies.

Solibri Model Checker (SMC) is a tool that, according to Solibri 2016, aids in finding and visualizing issues and problems both before and during construction. It also offers a wealth of information that can be utilized for a variety of purposes throughout the building's lifecycle, such as area calculations, accessibility, and building code compliance.

6. **BIM 360**: Autodesk's BIM 360 is a cloud-based platform that streamlines project management and collaboration for BIM processes. It includes components or modules for field management, document management, issue tracking, and design coordination. For project delivery and construction management, Autodesk designed an AEC cloud platform known as BIM 360; it comes in four different configurations and offerings: BIM Docs, BIM 360 Design, BIM 360 Glue, and BIM 360 Build (Khaerul 2023).

Figure 7: The four configurations of BIM 360 (Source: Khaerul 2023).

7. **Dynamo**: A visual programming tool for BIM software, like Revit, is known as Dynamo. It gives construction professionals and project stakeholders the ability to create their own scripts and automate repetitive tasks, enhance or improve productivity, and design exploration capabilities and discovery. The software works interactively with Revit for computational building information modeling (BIM) and does not work as a standalone product; Revit Dynamo is an incredibly powerful visual scripting plugin for Autodesk Revit that enables the average user to access the Revit API (Application programming interface) on an unprecedented level (Verma 2022).

8. **Bluebeam Revu**: The construction industry makes extensive use of the collaboration software or tool Bluebeam Revu, which works with PDF files. Construction papers can be marked up, annotated, and revision tracked, which

facilitates communication and collaboration on project designs or drawings and specifications.

9. **BIMobject**: A portal or platform known as BIMobject hosts a vast or extensive library of BIM objects or digital representations of building products and components. For use in their BIM models, design professionals can search for and download these objects.

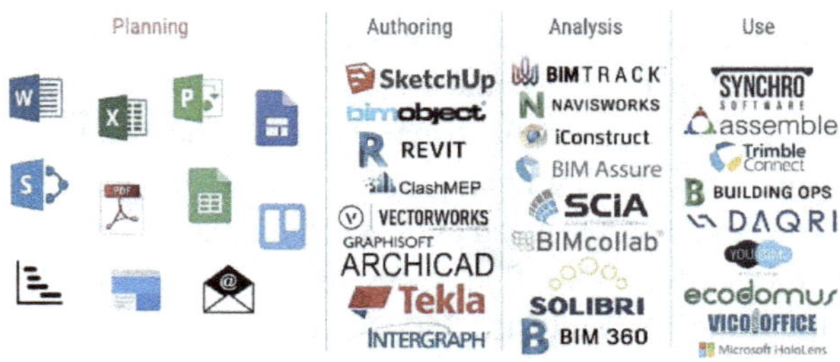

Figure 8: BIM planning, authoring, analysis and use tools (Source: Mirajkar 2021)

10. **Bentley AECOsim**: The Bentley Systems BIM platform AECOsim is designed specifically for projects in architecture, engineering, and construction. Collaborative workflows, clash detection, robust modeling capabilities, and quantity takeoff are all enabled. For a complete comprehensive design and analysis suite, AECOsim also integrates with other Bentley software applications, including MicroStation and OpenRoads.

11. **Vectorworks Architect**: A BIM platform that caters for architects and designers is known as Vectorworks Architect. It combines 3D modeling and advanced 2D drafting tools with BIM capabilities. Integrated workflows for architecture, landscape design, and entertainment design are all supported by Vectorworks Architect.

12. **Trimble SketchUp**: A popular or common 3D modeling software for interior and architectural design is called SketchUp. SketchUp offers BIM extensions, like Trimble Connect and Trimble Connect for HoloLens for collaboration and coordination, while not being typically considered or regarded as a full BIM platform.

13. **Nemetschek Allplan**: A BIM platform called Allplan was developed by Nemetschek Group. It offers thorough and comprehensive construction management, structural engineering, and architectural design tools. The core focus of Allplan is efficient workflows, design automation, and interdisciplinary collaboration.

14. **Graphisoft BIMx**: Graphisoft developed the mobile application BIMx. Users can navigate and explore interactive 3D models created in ArchiCAD with this tool. In order to improve communication and collaboration on construction sites, BIMx provides model visualization, data exploration, and augmented reality (AR) experiences. Like many other model viewers, it offers cross-references between 2D and 3D views and enables users to measure the separation between model points (Alison BIM course module 2023).

These are but a few illustrations of popular BIM software platforms and applications that are now on the market. Each platform has distinct strengths or advantages, features or capabilities, and user bases. The software choice will depend on the requirements of the project, team preferences, and industry standards or regulations.

3.1 Exploring the functionalities and capabilities of BIM tools

Building information modeling tools enables or gives a wide range of functionalities and capabilities that facilitate or enhance the planning, design, construction, and management of building projects. BIM tool functionality should make use of its technology to empower design and construction teams or experts to accomplish more in their respective fields while also having the ability to effectively integrate stakeholders throughout the collaboration phase of a BIM process. Also, a BIM tool capability should possess the basic ability to perform a specific task, deliver a service or generate a product in a BIM process. Below are some key functionalities and capabilities of BIM tools:

(A) **3D Modeling**: BIM tools enable the creation of highly detailed 3D models of buildings and infrastructure projects. These models include information about the physical geometry, materials, and components of the project. The 3D models serve as a digital representation of the project, allowing stakeholders to visualize and analyze the design.

(B) **Collaboration and Coordination**: BIM tools facilitate collaboration and coordination among project stakeholders. Multiple disciplines, such as architecture, structural engineering, and MEP (mechanical, electrical, and plumbing) systems, can work together within a shared BIM environment. This integration enables real-time coordination, clash detection, and resolution of conflicts early in the design process.

(C) **Information Management**: BIM tools act as centralized databases for project information. They store data about the building elements, such as specifications, performance characteristics, costs, and schedules. This information can be accessed and updated by different team members throughout the project lifecycle, ensuring data consistency and reducing errors.

(D) **Visualization and Simulation**: BIM tools allow for realistic visualization and simulation of the project. Users can generate high-quality renderings, animations, and walkthroughs to communicate design intent to clients and stakeholders. Additionally, BIM tools support various simulations, such as energy analysis, structural analysis, and clash detection, enabling performance evaluation and optimization.

(E) **Quantification and Cost Estimation**: BIM tools facilitate accurate quantification and cost estimation. The models contain information about the quantities and properties of building components, enabling automated takeoff and cost calculations. This streamlines the estimation process and enhances cost control throughout the project.

(F) **Construction Documentation**: BIM tools assist in generating construction documentation. They automate the creation of detailed drawings, schedules, and specifications from the BIM model. Changes made to the model automatically update the associated documentation, reducing manual rework and ensuring document consistency.

(G) **Facility Management**: BIM tools support the management of built assets throughout their lifecycle. BIM models serve as a repository of information for facility management tasks, including maintenance planning, asset tracking, and space management. This data can be accessed by facility managers to improve operational efficiency and reduce maintenance costs.

(H) **Integration with Other Software**: BIM tools often integrate with other software applications, such as structural analysis programs, cost estimation software, and project management tools. This integration enhances interoperability and data exchange between different software platforms, improving overall project efficiency and reducing data duplication.

These functionalities and capabilities of BIM tools contribute to improved collaboration, efficiency, accuracy, and decision-making throughout the entire lifecycle of a building project.

3.2 Choosing the right software for specific project requirements

Utilizing the appropriate BIM tools is important. When choosing a BIM tool or software, there are a number of things to keep in mind. Selecting the appropriate software for particular project needs or requirements is crucial to guarantee a seamless and successful integration or implementation of BIM processes. The following are some things to think about while choosing BIM software for a project:

a) **Project Scope and Complexity**: Evaluate the project's scale or size and complexities. Software with advanced modeling capabilities, clash detection, and coordination tools may be needed for larger and more complicated projects. Software that prioritizes user-friendliness or easy to use and efficient operations could prove advantageous for smaller projects.

b) **Professionals or Disciplines Involved:** Take into account the various disciplines or professionals working on the project, including structural engineering, architecture, MEP systems, and civil infrastructure requirements. Select software with tools and features that are specific to those fields or disciplines.

c) **Collaboration and Interoperability**: Evaluate how well the program collaborates and how well it can exchange data with other software platforms. Make sure it works with the standards and file formats that the project's stakeholders prefer. Software that facilitates cloud-based collaboration should be taken into consideration for easy data sharing and communication.

d) **User Experience and Learning Curve**: Study the software's ease of use and learning curve. Take into account the members of the project team's skill levels and level of BIM tool expertise. Select software that offers sufficient tutorials, training materials, and support to aid in the team's adoption and efficient use of the software program.

e) **Integration with Existing Workflows**: Assess the degree to which the selected software can be integrated with the project team's current software applications and workflows. Productivity is enhanced, procedures are streamlined, and there is less need for manual data transfers due to the seamless integration or connection.

f) **Customization and Extensibility**: Selecting a BIM tool or software for a project's undertakings, think about whether the program supports scripting, APIs, or plugins for customization and extension (Moreno et al. 2022). This can be useful for automating tedious tasks or customizing the software to meet project requirements.

g) **Vendor Support and Updates:** Examining the reputation and level of support provided by the software provider is a crucial step in the selection of BIM software. Take into account factors including the software's long-term development roadmap, the vendor's reaction to issues, and the availability of updates and bug fixes.

h) **Licensing and Budget**: Take into account the project's total budget and plan, the software's licensing alternatives, and its pricing model. Determine carefully if the software's price is in line with the needs or requirements of the project and off course the expected return on investment.

Key stakeholders, including owners, contractors, engineers, and architects, must be included in the decision-making process to guarantee that the software selected satisfies their unique requirements and advances the project's objectives. Furthermore, using the software that has been shortlisted for pilot projects or trials might yield important information on how well it fits the project's requirements or needs.

Chapter 4: BIM Workflow and Collaboration

A Building Information Modeling (BIM) workflow is a process that makes use of BIM to enhance project design, construction, and operation; the process integrates tools or software to create a unified paradigm that enables teams to collaborate on projects at every stage. Workflow and collaboration associated with BIM (Building Information Modeling) are essential in the construction industry. A building's or infrastructure's functional and physical characteristics are represented digitally using BIM. Throughout the course of the project, it allows stakeholders to create, manage, and exchange information.

The BIM workflow consists of several stages, including:

	BIM Workflow Stages	*Stages Description*
1	Conceptualization and Design	Using BIM software, designers and architects can create a virtual model of the building that includes its geometry, spatial relationships, and various components. Better design options analysis, visualization, and evaluation are made possible by BIM.
2	Documentation and Detailing	BIM models are further developed, and precise details are included, including materials, quantities, and specifications. Accurate construction documentation, such as schedules, specifications, and drawings, can be generated with the help of this information.
3	Coordination and Clash Detection	Integrating or merging models together from various disciplines, including architectural, structural engineering, and MEP (mechanical, electrical, and plumbing) systems, allows BIM to facilitate multidisciplinary coordination. Clash detection technologies reduce errors and rework during construction by identifying and resolving conflicts between different building systems.
4	Construction and Project Management	Construction planning and sequencing are made easier with the use of BIM models. They can be used for scheduling, cost estimation, and quantity takeoffs. Better project management and decision-making can result from better stakeholder collaboration and communication made possible by BIM.
5	Operation and Maintenance	BIM models are useful beyond the construction stage. They offer facility managers and owners a wealth of

		information. Throughout the lifecycle of the building, BIM can be integrated with asset management systems to facilitate effective maintenance planning, asset tracking, and performance analysis.

Table 3: BIM Workflow Stages

Several disciplines work together during the BIM process to create a digital representation of a building's physical and functional characteristics. A successful workflow in BIM requires effective collaboration. It involves the following aspects:

	BIM Collaboration Workflow	*Description*
1	Shared Data Environment	This makes it feasible for collaborators to access and exchange BIM models, documentation, and project-related data within a common working environment or platform. A reliable source of up-to-date or latest and accurate information is ensured for all stakeholders involved by this shared data environment.
2	Version Control and Change Management	BIM models change throughout time as a result of numerous contributors' updates and modifications. Version control systems or mechanisms facilitate the monitoring of modifications and maintain the history of model revisions or iterations. Modifications are appropriately evaluated, approved, and communicated to all team members due to change management processes.
3	Collaborative Workflows	Collaborators effectively coordinate their contributions by adhering to an established standard workflow. The identification and resolution of conflicts as well as the smooth operation of disciplinary collaboration are facilitated by BIM coordination meetings, clash detection sessions, and frequent information exchanges.
4	Communication and Information Sharing	Effective collaboration requires clear channels of communication. Information, concerns, and updates should be shared between collaborators in a timely and organized manner. To facilitate communication, BIM collaboration tools often incorporate features like comments, markups, and real-time chat.

5	Training and Education	Project stakeholders or team members should receive training on BIM workflows, processes, and tools to ensure successful collaboration. This ensures that every person concerned understand the fundamentals of BIM and can make valuable contributions to the collaborative endeavors.

Table 4: BIM Collaboration Workflow

To sum it up, better communication, coordination, and decision-making are made possible throughout the lifecycle of a construction project due to BIM workflow and collaboration. Through the utilization of BIM technology and the application of effective collaboration approaches, stakeholders can optimize efficiency, minimize errors, and deliver buildings of higher standard or quality.

4.1 Understanding the collaborative nature of BIM

Building Information Modeling is by its very nature collaborative, bringing together multiple stakeholders involved in design, construction, and the operation of a building or infrastructure project. The collaborative aspect of the project arises from the capacity to generate a common digital model that functions as a central information repository, facilitating efficient communication and coordination amongst the participants in the project.

These are some key aspects that demonstrate how collaborative BIM is:

A. **Shared Information**: BIM makes it easier to create and share a single, comprehensive model that incorporates information from many disciplines. A thorough or comprehensive understanding of the project can be ensured by including the knowledge and experience of architects, engineers, contractors, subcontractors, and other stakeholders in the model. By establishing a common understanding and a base for decision-making, this shared information enhances collaboration.

B. **Multidisciplinary Coordination:** Stakeholders from many disciplines can collaborate easily through BIM. Architects, structural engineers, MEP (mechanical, electrical, plumbing) designers, and other professionals can discover and resolve conflicts or clashes between systems early in the design phase by using the model as a platform for interdisciplinary collaboration. This collaborative coordination reduces errors, lowers rework, and optimizes project efficiency overall.

C. **Enhanced Communication**: Effective communication between project participants is facilitated by BIM. Within the BIM software environment, stakeholders can exchange information, annotate the model, and give comments. Real-time collaboration facilitates swift issue resolution, better team alignment, and clarification of design purpose. BIM-supported clear communication channels foster collaboration, understanding, and the development of a common project vision.

D. **Simultaneous Workflows**: BIM enables multiple stakeholders to work together on different project aspects at the same time. For instance, architects can design the geometry of a building, structural engineers can assess the same building's structural integrity, and MEP designers can plan and set up the building systems. The collaborative nature of BIM allows stakeholders to access and modify the model concurrently, ensuring effective progress and integration of their work. This allows for these simultaneous workflows.

E. **Data Exchange and Integration**: BIM makes it easier for data to be integrated and exchanged between different disciplinary software programs used by different stakeholders. Information from different sources, including design software, cost estimation tools, and scheduling software, can be easily integrated into the BIM model due to open standards and interoperability. By giving stakeholders, a uniform view of project information and empowering them to make informed decisions based on correct or accurate and current or up to date data, this data integration supports collaboration.

F. **Lifecycle Collaboration**: Beyond the construction stage, during the building's asset's operation and maintenance, BIM promotes collaboration. Facility managers can receive the BIM model and use it for asset tracking, maintenance planning, and space management. By collaboration, the design and operations teams can optimize the building's performance and efficiency throughout the building or facility's lifecycle and guarantee a seamless transition from construction to facility management.

All things considered, BIM's collaborative nature stem from its capacity for bringing together stakeholders, promote data exchange and integration, enable multidisciplinary coordination, facilitate information sharing, improve communication, support concurrent or simultaneous workflows, and improve collaboration throughout the project lifecycle. Stakeholders may fully utilize BIM to deliver successful projects with more efficiency, reduced costs, and higher quality outcomes by adopting this collaborative approach.

4.2 Examining the BIM workflow process from design to construction.

A smooth transition from the design to the construction phases is ensured by a number of steps and activities in the BIM (Building Information Modeling) workflow process. This is an overview of the common steps that are involved.

4.2.1. Design Phase

a) **Conceptualization Stage**: Initial design concepts and sketches are created by the design team. Project stakeholders can see the design by using BIM software such as "Revit" to generate a 3D digital model of the building.

b) **Detailed Design Stage**: The design at this stage is further refined, and the BIM model gets more details on structural systems, architectural elements, MEP (mechanical, electrical, and plumbing) systems, and other features. Different design disciplines collaborate in order to ensure coordination and detection of clashes.

4.2.2. Quantity Takeoff and Cost Estimation Phase

a) Quantities of components and materials are extracted for procurement and cost estimation using the BIM model. This process can be automated with BIM software, saving time and reducing errors.

b) In order to generate accurate cost reports for financial planning and budgeting, cost estimating is performed based on the quantities.

4.2.3. Construction Planning Phase

a) **Sequencing and Scheduling**: Project teams can create sequence activities and identify critical paths by using BIM to produce a construction schedule. Visualizing the construction sequence and identifying dependencies are made easier by the BIM model.

b) **Logistics Planning**: Analyzing site logistics, such as material storage, access routes, and crane locations or placements, can be aided by BIM.

By doing so, construction operations and processes are optimized, and disruptions are reduced.

4.2.4. Procurement and Prefabrication

a) **Material Procurement**: Accurate quantities and specifications are provided by BIM models, which facilitate the procurement process. Supply chain management system integration can streamline and speed up material ordering and tracking.

b) **Prefabrication and Off-Site Construction**: The planning and coordinating of prefabrication and off-site construction operations or activities are facilitated by BIM models. The model's accurate data makes it possible to produce and assemble prefabricated components quickly and effectively.

4.2.5. Construction Execution

a) **On-Site Coordination**: On-site, BIM models are used to facilitate coordination between various contractors and trades. Understanding design intent, resolving clashes or conflicts, and coordinating the installation of various building systems are all made easier with the help of the model.

b) **Construction Monitoring**: To maintain and monitor site work progress, quality, and safety, BIM can be integrated with sensors and data collecting technologies or devices on the construction site. The construction process can be effectively managed and controlled by comparing real-time data from the site with the BIM model.

4.2.6. As-Built Documentation

a) The BIM model is updated to show or reflect the actual building conditions as construction work progresses. This as-built model turns into a valuable tool or asset for maintenance and operations in the future.

b) The BIM model generates documentation, such as as-built drawings, operation and maintenance manuals, and asset information, providing and ensuring accurate and thorough or comprehensive data for facility management.

BIM makes it possible for project stakeholders to collaborate, visualize, coordinate, and exchange data throughout the design and construction phases. The BIM model acts as a central information repository, facilitating a streamlined workflow process that enhances efficiency, reduces errors, and boosts project outcomes.

4.3 Collaborative platforms and coordination techniques in BIM

As previously established, the Building Information Modeling workflow method is collaborative in nature. In order to collaborate, exchange information, and work simultaneously on a common shared platform, stakeholders in construction projects will require a platform. Effective communication, information sharing, and coordination among stakeholders in a BIM (Building Information Modeling) environment are made possible in large part by collaborative platforms and coordination strategies or techniques. Below are some of the popular or common platforms and techniques used in BIM workflow process:

	Types of collaborative platforms	Platforms details explanation
1	Common Data Environment (CDE)	Project stakeholders can access and exchange BIM models, documents, and other project-related data on a centrally located digital platform called a CDE. It ensures that everyone is working with the most recent and accurate data by offering a single source of truth. To efficiently handle collaboration, CDEs frequently incorporate features like version control, permissions for access, and audit trails.
2	BIM Collaboration Software	A variety of software tools offering features such as real-time model viewing, markups, comments, clash detection, and issue tracking, are specifically designed for BIM collaboration. Regardless of where the project team are physically located, stakeholders can work collaboratively directly within the BIM model due to these capabilities. Graphisoft BIMcloud, Trimble Connect, and Autodesk BIM 360 are a few examples of BIM collaboration software.
3	Clash Detection and Coordination Meetings	Project stakeholders can use clash detection software to identify and resolve conflicts or clashes between various building systems (such as structural, architectural, and MEP systems) inside the BIM model. Multidisciplinary teams meet for coordination meetings in order review clash reports, discuss solutions or resolutions, and to ensure that different disciplines are coordinated. These kinds of meetings are necessary for effective collaboration and conflict resolution.
4	Model Federations	The process of merging or integrating and coordinating BIM models from different disciplines into a single, coordinated model is known as model

		federations. Stakeholders can visualize the whole building design through federations, which could assist them to identify coordination issues and make required corrections or adjustments. They ensure that the various models are synchronized or coordinated and aligned, reducing clashes and inconsistencies or discrepancies during construction.
5	Open BIM and Industry Standards	The term "open BIM" describes the implementation of open standards and file formats that facilitate software application interoperability among various stakeholders. By enabling exchange and integration of BIM data across different platforms, open BIM fosters collaboration. BIM information may be shared and transferred more easily by means of standards like Industry Foundation Classes (IFC) and Construction Operations Building Information Exchange (COBie).
6	Virtual Design and Construction (VDC)	3D, 4D (time), and 5D (cost) modeling are used in VDC techniques to enhance coordination and collaboration. From the BIM model, stakeholders can extract construction-related data, analyze the effects of construction, and simulate and display construction sequences. By identifying potential issues, streamlining, and optimizing the construction process, VDC techniques enhance collaboration and outcomes for the project.
7	Cloud-Based Collaboration	Project stakeholders can access collaboration tools and BIM models from any location with an internet connection due to cloud computing. Real-time synchronization is made possible by cloud-based collaboration platforms, which let multiple stakeholders access and modify BIM models simultaneously. This fosters collaboration, particularly when project team members are located far apart geographically.
8	Mobile Applications	Using their mobile devices, project stakeholders can access, view, and review BIM models through mobile apps designed for BIM collaboration. This makes field inspections, issue reporting, and on-site coordination easier. Mobile apps promote real-time coordination and collaboration by providing stakeholders quick access to project information.

Table 5: Types of BIM collaborative platform (Source: author)

By employing these collaborative platforms and coordination methodologies, stakeholders can boost communication, speed information exchange, identify and resolve disputes, and ensure optimal coordination throughout the BIM workflow. These resources, tools, and techniques enhance collaboration, minimize rework, and result in better outcomes for projects.

Chapter 5: BIM and the Design Process

Building information modeling is the process of creating a virtual model of a building that contains details on its geometry, materials, systems, components, and other elements using computer-aided design (CAD) software and other technologies. The BIM process begins with the use of CAD software and other tools for developing a virtual model, much like any other traditional design metho.

Building Information Modeling (BIM) has brought about a collaborative and integrated approach to building design and construction, which revolutionized the design process in the architecture, engineering, and construction sectors. This is how BIM influences the process of design.

1. **Design Visualization**: Using BIM, designers or architects can produce three-dimensional (3D) models of buildings that give a realistic visual representation of the project. This facilitates better understanding of the design intent and informed decision-making for all stakeholders involved, including clients and designers.

2. **Design Coordination**: Better coordination and integration of design elements is made possible by BIM, which makes it easier for professionals from various design disciplines, such as Architects, structural engineers, mechanical engineers, and other professionals, to collaborate or coordinate on a shared BIM model. This reduces conflicts and clashes between different systems.

3. **Design Data Integration**: The integration of various data types within the model is made possible by BIM. It can include information regarding construction schedules, materials, specifications, and cost estimations, among other things. The accuracy as well as efficiency of the design process is improved by this integrated approach.

4. **Design Analysis and Simulation**: Designers can perform a variety of simulations and analysis on the model through BIM. An energy analysis, for instance, can be carried out to optimize the building's energy efficiency. Also, structural evaluation or analysis can be carried out to assess the building's structural integrity. These evaluations support the early detection or identification and resolution of possible or potential issues during the design phase or process.

5. **Automate Iterative Design**: An iterative design method is supported by BIM. Designers can change the model, and those modifications or changes are automatically

and immediately reflected across all related drawings and components. This reduces errors and saves time by streamlining the design iteration process.

6. **Design Collaboration**: BIM facilitates collaboration among members of the project team. The BIM model can be accessed, and work contributions carried out by multiple stakeholders simultaneously. Better design outcomes come about due to improved communication, coordination, and decision-making.

7.**Construction Documentation**: The generation of construction documentation is made easier by BIM. The time and effort needed for manual documentation can be reduced because the BIM model can automatically generate 2D drawings, schedules, and specifications since it contains accurate and up-to-date information.

8. **Facility Management**: The advantages of BIM go beyond the design and construction stages. When it comes to facilities management, the BIM model is a valuable tool or resource. It offers thorough or comprehensive information about the building's equipment, systems, and maintenance requirements, aiding in the facility's effective operation and maintenance.

In a nutshell, BIM improves design visualization, coordination, data integration, analysis, collaboration, and documentation, all of which have a significant influence on the design process. Throughout the entire lifecycle of a building project, BIM enhances efficiency, accuracy, and collaboration, leading to better-designed structures and better project outcomes.

5.1 Integrating BIM into the architectural design phase.

Building Information Modelling can be fully integrated into the architectural design stage of a project. Architects are being compelled to continuously improve their architectural design techniques due to the rapid development of BIM integration with emerging technology (Abdelhai 2022). BIM has several benefits and can enhance the architectural design process significantly. Below are some essential aspects of integrating BIM into the stage of architectural design.

1. **Conceptual Design**: From the initial phase of conceptual design, BIM can be used. Architects are able to visualize and comprehend the proposed building more clearly by using 3D models of their design concepts. This facilitates the exploration of various design possibilities and the successful communication of the design intention.

2. **Design Development**: Architects can incorporate more detailed information into their 3D model as the design develops through BIM. They can include

components, systems, materials, and other design elements to produce a thorough virtual representation or model of the structure.

3. **Visualization and Presentation:** With the use of BIM's 3D modeling features, architects can generate accurate visual representations of their building designs. The design can be communicated to clients, stakeholders, and the project team using these representations. Additionally, BIM makes it easier to create flythroughs and walkthroughs, giving users an immersive experience of the design.

4. **Design Coordination**: Better coordination between structural engineering, mechanical systems, and architectural design is promoted by BIM. Architects can reduce the need for costly modifications or changes during construction by identifying and resolving conflicts and clashes early in the design phase by using a shared BIM model.

5. **Design Analysis**: Architects can use BIM's various design analysis capabilities or tools to optimize their designs. Energy analysis tools, for instance, can be used to assess the building's energy performance and pinpoint or identify areas in which energy efficiency might be improved. Natural lighting conditions can be adjusted with the aid of daylight analysis. Through the implementation or conduction of such analyses within the BIM environment, architects can make informed design decisions driven by data.

6. **Parametric Design**: Parametric modeling features are often found in BIM tools. This enables architects to produce or create intelligently designed objects that can be parameter manipulated and controlled. Architects may quickly explore design variations, study the impact of design modifications, and expedite the iterative design process with parametric design.

7. **Quantification and Cost Estimation**: Costs, quantities, and material information can all be stored in BIM models. Throughout the design process, architects can use this data to produce precise quantity takeoffs and cost estimates that will help with budgeting and cost control.

8. **Collaboration and Communication**: The project team members collaborate more when they use BIM. Architects may effectively facilitate coordination and communication by sharing the BIM model with engineers, consultants, contractors, and other project stakeholders. A more efficient project delivery process and better design integration result from this collaborative approach.

9. **Construction Documentation**: The creation of construction documents is made easier by BIM. Architects can automatically create construction drawings, schedules, and specifications since the BIM model has precise and up-to-date data. This ensures consistency and lowers documentation errors.

Architects stand to gain from enhanced visualization, coordination, analysis, collaboration, and documentation when BIM is incorporated into the architectural design phase. BIM improves architectural design's effectiveness, precision, and quality, which eventually results in better-designed structures or buildings.

5.2 Using Building Information Modelling for Spatial analysis, Visualization, and Simulation

Building Information Modeling provides strong tools for spatial analysis, visualization, and simulation, architects and designers may make well-informed decisions and obtain insightful information. Finding the locations of geographic features and phenomena, as well as their spatial relationships or interactions with one another can be done with the help of spatial analysis (Hao 2019). BIM can also be used to test construction sequences, simulate alternative design possibilities, and assess how well building systems are performing. Visualization is the means through which stakeholders can immediately view and access these actions. Below is how BIM can be used for these purposes or goals.

1. **Spatial Analysis**: Architects can evaluate and optimize a building's spatial layout with the aid of BIM's spatial analysis tools. For instance, within the model, architects can analyze accessibility, space usage, and circulation patterns. They can evaluate how well space is being allocated, spot possible bottlenecks, and make adjustments or changes to enhance functionality and user experience.

2. **Daylight and Sunlight Analysis**: Features for evaluating daytime and sunlight conditions are frequently found in BIM tools. The way that natural light will interact with a building at different times of the day and year can be simulated and visualized by architects. This analysis aids in placing windows, skylights, and shading devices optimally to maximize daylight penetration, lower energy consumption, and create convenient or comfortable interior environments.

3. **Thermal Analysis**: Thermal analysis and simulation through BIM can be used to evaluate the energy performance of a building. Within the BIM environment, architects can model the building envelope, materials, and HVAC (Heating, Ventilation, and Air Conditioning) systems. They are able to assess heat gain and loss,

thermal comfort, and energy usage as a result. It assists in determining energy-saving tactics, such enhancing insulation or optimizing HVAC systems.

4. **Energy Analysis**: Through the integration of energy simulation software, BIM facilitates energy analysis. Architects are able to assess or evaluate the energy efficiency of the building design, pinpoint possible energy-saving strategies, and calculate carbon emissions and energy use. Making informed decisions about renewable energy integration, energy-efficient design techniques, and optimizing building performance is made easier by this analysis.

5. **Acoustic Analysis**: Acoustic properties and acoustic analysis can be supported through BIM models. Architects are able to simulate and assess noise levels, acoustic quality, and sound transmission in different spaces. This analysis helps in mitigating noise-related problems or issues, designing suitable acoustic treatments, and optimizing space or room layouts for desired acoustic performance.

6. **Structural Analysis**: Architects can conduct or perform structural simulations and analyze the behavior of the building's structural elements by integrating BIM models with structural analysis tools or software. They are able to assess the response to different loads and forces, load-bearing capacity, and structural integrity. This analysis helps to maximize or optimize the design for structural safety and efficiency as well as to identify possible structural problems or issues.

7. **Visualization in 4D and 5D**: Architects are able to visualize and simulate the construction process (4D) and related costs (5D) with the use of BIM, which can incorporate time- and cost-based information. Construction sequencing visualizations, which depict the project's progression across time, can be created by architects. Based on the BIM model, they may also create project schedules and estimate costs. Planning, coordinating, and communicating with stakeholders are made easier by these visualizations and simulations of projects.

Architects can make data-driven design decisions, maximize building performance, improve user experience, and improve project outcomes by using BIM for spatial analysis, visualization, and simulation. A comprehensive understanding of the building's performance is provided by BIM's capacity or potential to integrate various or multiple analyses within a single model, which empowers architects to create more effective, sustainable, and practical or functional designs.

5.3 Exploring design options and optimizing performance in BIM

Through the use of performance simulation tools, parametric building generative models, and optimization algorithms, the design optimization process can identify high-performing design variants in performance-based architecture design by exploring a range of options (Liu, Wang, and Ji 2022). Architects may explore design alternatives or options and optimize or enhance performance with the use of Building Information Modeling (BIM), leading to more effective and efficient design results. This is how BIM may be used by architects for these kinds of purposes.

A. **Design Optioneering**: Architects can quickly create and evaluate many or multiple design options in a virtual environment through BIM. Architects may explore diverse space or spatial arrangements, material choices or selections, and architectural elements by developing alternate designs and variations. The parametric modeling features of BIM make it possible to manipulate design elements effectively, which makes it easier to explore different design iterations.

B. **Performance Analysis**: Architects can assess the performance of various design options with the aid of performance analysis tools integrated into BIM. Energy analysis tools, for instance, may simulate and assess the energy efficiency of alternative design options. In order to determine the best design in terms of energy performance, architects can compare the sustainability and energy consumption metrics of each alternative.

C. **Iterative Design**: Iterative design is facilitated by BIM, allowing architects to modify the model and evaluate how their modifications affect different performance parameters or metrics. Architects can optimize various factors like energy efficiency, daylighting, thermal comfort, structural integrity, and acoustics by iteratively refining the design based on analysis results.

D. **Cost Optimization**: Architects can take cost optimization into account while designing using BIM. Architects can evaluate the cost implications of various design choices or decisions by integrating cost data and doing cost analysis within the BIM environment. In order to optimize cost efficiency, they might weigh or evaluate the trade-offs of design options, taking into account things like material choices, construction techniques, and building systems.

E. **Sustainability Integration**: The architectural process can more easily include sustainable design principles with the aid of BIM. Using BIM, architects could incorporate sustainable features like rainwater harvesting, green roofs, and renewable energy systems. Architects can evaluate the effects or impacts of these sustainable elements on energy efficiency, carbon emissions, and sustainability in general by simulating their performance.

F. **Collaboration and Stakeholder Engagement**: In the design exploration stage, BIM encourages collaboration and stakeholder involvement. Architects can solicit input and feedback from clients, consultants, and other stakeholders by sharing the BIM model with them. This collaborative method ensures that design decisions correspond or align with the objectives, requirements, and specifications of the project.

G. **Design Optimization and Refinement**: The visualization capabilities of BIM enable architects to assess the functional and aesthetic aspects of various design options. To better grasp the spatial qualities or characteristics and user experience connected to different design alternatives, they can produce realistic renderings or depictions, walkthroughs, and virtual reality experiences. This aids in improving the design and making informed decisions based on feedback from the visuals.

Architects may produce more effective, sustainable, and aesthetically appealing designs by utilizing BIM to investigate design options and optimize performance. By combining analysis tools, collaboration features, and visualization capabilities, BIM enables architects to maximize design performance, make data-driven decisions, and produce high-quality architectural solutions.

Chapter 6: BIM and Construction Management

Building or infrastructure projects require a detailed design involving architects, engineers, and other pertinent stakeholders for success. This is followed by construction operations. Construction management is the technical term describing the necessity for appropriate on- and off-site project management. Building information modeling (BIM), and construction management are two related disciplines that are essential to the construction industry.

Construction management involves overseeing and coordinating every facet of construction, from the very beginning of a project to its completion. Plans, organizing, and budget allocation and management of the project's finances, time, and resources fall within the purview of construction managers. They ensure the construction process aligns with safety rules, quality standards, and project objectives. Construction managers' primary responsibilities consist of:

	Responsibilities	*Explanations*
1	Project Planning	Project goals, schedules, budgets, and resource requirements are all specified in the comprehensive project plans that construction managers prepare.
2	Procurement	They manage the entire procurement process, including supplier, subcontractor, and contractor source and selection.
3	Resource Management	Construction managers make sure labour, supplies, and machinery and equipment are accessible when needed by allocating resources effectively.
4	Risk Management	They identify potential risks and create plans or strategies to mitigate them to keep the workplace safe and secure.
5	Quality Control	Construction managers oversee the construction process, guaranteeing compliance to regulations, codes, and specifications while maintaining quality standards.

Table 6: Construction Managers Duties (Source Author)

All these duties carried out by a construction manager during the construction process require interacting with or collaborating with project stakeholders. Communication and better collaboration between all relevant stakeholder engaged in the construction process, including architects, engineers, contractors, owners, and other agents, are made easier by BIM. Advantages of BIM for construction projects include:

	Advantage	Detail
1	Enhanced Visualization	BIM enabling stakeholders to see the project in 3D or even 4D (by incorporating the time dimension dimension) models, facilitating decision-making and design comprehension.
2	Improved Coordination	Clash detection is made easier by BIM, which reduces rework and saves time and money by spotting conflicts between various building systems (such as HVAC and electrical) before construction begins.
3	Efficient Project Planning	Accurate quantity takeoffs, cost estimates, and scheduling are made possible by BIM, which also optimizes resource allocation and guarantees project completion on time.
4	Enhanced Collaboration	By offering a single or a consolidated platform for sharing and accessing project information, BIM helps project teams collaborate by promoting improved coordination and communication.

Table 7: Advantages of BIM to construction projects (Source: author)

A. Improve Planning: Construction managers may better plan and make decisions by using BIM to visualize and simulate the project.

B. Enhance Coordination: Building Information Modeling (BIM) enables construction managers to spot and address issues at an early stage of the design process, thereby avoiding expensive rework during construction.

C. Streamline Communication: Construction managers can share and access data with project stakeholders through the use of BIM, which serves as a centralized hub or location for project information, enhancing communication and collaboration.

D. Optimize Resource Allocation: Construction managers may effectively manage resources and procurement by using BIM's quantity takeoff capabilities to estimate materials and resources with accuracy.

E. Enhance Project Control: With the use of real-time data and progress tracking provided by BIM, construction managers can keep an eye on the project's performance and make sure that timetables and budgets are followed.

Construction management and building information modeling (BIM) are closely related. BIM as a valuable tool helps construction managers plan, coordinate, communicate, and control projects more efficiently, which improves project outcomes.

6.1 Applying BIM for construction sequencing and scheduling (BIM-4D)

In the construction sector, there are several benefits of using BIM (Building Information Modeling) for sequencing and scheduling construction projects. Construction processes may be visualized, analyzed, and optimized with the help of BIM, which makes project activity sequencing and scheduling more effective.

Construction managers can produce 3D models of the project with BIM, integrating all pertinent information about the systems, components, and building elements. Stakeholders can obtain a full picture of how various operations interact and progress over time by visualizing the construction process in a virtual environment.

Construction sequencing and scheduling are made possible with 4D BIM by incorporating a time dimension into the 3D BIM model. This means linking project activities to their corresponding timelines in order create a time-integrated model that shows the construction sequence. Through the visualization and analysis of the construction process, 4D BIM assists stakeholders in spotting possible conflicts, clashes, and inefficiencies.

Clash detection, which finds conflicts or clashes between various building elements or systems that might occur during construction, is made possible by BIM software. Construction managers can prevent expensive rework and delays by proactively resolving conflicts in the virtual model before they come on actual construction site.

Construction managers can simulate and analyze diverse construction sequences and scenarios with the use of BIM software. Construction managers can determine which sequencing options are the most efficient and time-saving by virtually testing and optimizing them. Making informed decisions about construction techniques, resource allocation, and logistical planning is aided by this approach.

With the help of precise quantity takeoffs from BIM models, construction managers can calculate the amount of labor, materials, and equipment needed for each task. The scheduling information and this data can be integrated to create comprehensive resource plans. Construction managers can prevent resource shortages, optimize material deliveries, and guarantee timely completion of operations by matching quantities and resources with the construction schedule.

BIM acts as a platform for collaboration, facilitating stakeholders in a project's coordination and communication. The project team, including suppliers and subcontractors, can all clearly comprehend the construction sequencing and schedule when the BIM model is shared with them. This enhances overall project coordination, fosters effective collaboration, and minimizes misunderstanding.

Real-time tracking and monitoring of construction activity progress is possible using BIM. Construction managers can quickly identify any deviations from the intended schedule by comparing the actual progress made on-site with the BIM model. This makes it possible to have more control over the construction process and to make timely modifications to keep the project on schedule.

In a nutshell, the utilization of Building Information Modeling (BIM) in construction sequencing and scheduling leads to improved project planning, coordination, and efficiency. Construction managers may optimize construction sequences, minimize conflicts and delays, enhance resource planning, and guarantee successful project completion within the allotted timeframes by utilizing the visualization, clash detection, simulation, and analysis features of BIM.

6.2 BIM for project Estimation and Cost Modelling (BIM-5D)

In the construction industry, building information modeling, is a useful tool for project estimation and cost modeling. Construction experts may precisely estimate project costs, develop comprehensive cost models, and determine on resource allocation and budgeting by utilizing the data-rich BIM model.

Accurate quantity takeoff, or the extraction of material, component, and system quantities straight from the BIM model, is made possible by BIM. BIM software may automatically calculate amounts by assigning particular properties and attributes to each element in the model. By doing this, the estimation process is streamlined and the possibility of errors in manual quantity takeoff methods is reduced.

Also, cost information can be incorporated into the model by using BIM. Each building element, system, or component in the BIM model can have cost information assigned to it by construction professionals. This includes unit costs, labor rates, material costs, and equipment costs. Accurate cost estimates based on the quantities taken out of the model can be carried out by connecting cost data with the BIM model.

Parametric cost modeling, in which cost data is connected to particular parameters or attributes in the BIM model, is made easier by BIM. Cost modeling can be completed quickly and dynamically by adjusting some parameters, such as the area, length, or height of the elements, and have the associated costs immediately and automatically recalculated. This makes it possible for professionals in the construction industry to assess various design alternatives and comprehend the financial effects or cost implications of design choices.

In addition, Visual representations of cost data are provided by BIM, which improves cost modeling. Construction professionals can examine project-wide cost distributions using the BIM model, highlighting areas of high-cost concentration or possible cost savings. This visual cost analysis facilitates the comprehension of cost consequences, the optimization of design decisions, and the successful dissemination of cost-related information to stakeholders.

Throughout the project lifecycle, real-time cost tracking and control are supported by BIM. Construction professionals can track and update project costs in real-time by integrating the BIM model with software for project scheduling and cost management. This allows for better cost control, identifying any deviations from the budget and enabling proactive measures to address cost overruns or savings opportunities.

Project teams can work together on cost estimation using BIM as a collaborative platform. Stakeholders like architects, engineers, contractors, and estimators can work together to produce accurate estimates by sharing the BIM model, which gives everyone access to the same cost-related data. This guarantees a shared knowledge of project costs, enhances communication, and lessens disagreements or conflicts in cost estimation.

The estimating software and the BIM model can exchange data seamlessly when the estimating software and BIM are connected. This integration makes it easier to model costs in more detail and sophistication by utilizing the sophisticated features and functionalities of estimating software and the wealth of data present in the BIM model.

BIM offers a strong foundation for cost modeling and project estimation. BIM improves the precision, efficacy, and efficiency of cost estimation and modeling procedures in construction projects by enabling accurate quantity takeoff, integrating cost data, supporting parametric cost modeling, facilitating visual cost analysis, enabling real-time cost tracking and control, promoting collaboration, and integrating with estimating software.

6.3 Utilizing BIM for clash detection and coordination

In the construction industry, one of the fundamental applications of BIM (Building Information Modeling) is for clash detection and coordination. During the construction process, different building elements, systems, or components may clash or have conflicts. BIM offers a strong platform to identify and manage these conflicts.

Building Information Modeling involves creating a digital model that contains all project information, including MEP (mechanical, electrical, and plumbing), structural and architectural systems. By functioning as a single data repository, this model enables stakeholders to access collaborate simultaneously on a comprehensive representation of the project.

Collisions can occur between architectural and structural components, MEP clashes, or elements of different trades. BIM software offers automated clash detection capabilities that identify potential conflicts or clashes that may occur when different elements overlap or intersect in the model. Clash detection algorithms compare the spatial relationships between elements and generate clash reports, highlighting the locations and types of clashes.

The positive aspect is that BIM's clash detection feature makes it possible to find conflicts early on, before construction ever starts, during the design and planning phases. Construction teams can prevent costly delays and rework by proactively resolving conflicts with virtual clash detection during the construction phase.

Building Information Modelling serves as a platform for collaboration, enabling efficient coordination and communication amongst project stakeholders. Architects, engineers, contractors, and subcontractors can see and comprehend clashes by sharing clash detection reports that are produced from the BIM model. This encourages collaboration and enables stakeholders involved to come up with solutions and make plausible decisions to resolve the conflicts.

By providing stakeholders access to tools and features that facilitate effective conflict resolution, BIM aids in the resolution of conflicts. Stakeholders can evaluate and analyze clashes in the 3D model, exploring different options and solutions, based on clash detection results. To avoid conflicts, this procedure might involve adjusting designs, moving elements elsewhere, or rearranging construction activities. Clashes can be eliminated by iterative refinement, ensuring a construction process free of conflicts.

Clash-free coordination reduces on-site conflicts, minimizes rework, and improves overall productivity, leading to smoother construction operations. BIM's clash detection capabilities improve coordination among various disciplines and trades. Construction teams can better coordinate their activities, aligning schedules, materials, and installation sequences, by identifying clashes early on.

Also, beyond clash detection in isolated components, BIM can also detect clashes in construction sequences. Clash detection on the temporal relationships between activities and their associated elements can be carried out by incorporating the time dimension (4D BIM). This facilitates the identification of construction sequencing conflicts and the creation of optimal construction schedules by stakeholders.

There are several advantages to using BIM for clash detection and coordination in construction projects. These advantages include improved coordination between project stakeholders, a reduction in clashes and rework, increased construction efficiency, and successful project outcomes. BIM made this possible by its comprehensive digital model, clash detection tools, collaboration features, and early conflict resolution abilities.

6.4 BIM's role in construction documentation and quality control

Construction documentation and quality control measures heavily rely on BIM (Building Information Modeling). It offers a digital platform that makes it easier to create, maintain, and use construction documents and makes it possible to implement efficient quality control measures.

Building Information Modeling acts as a central repository for project data, making it easier to create precise and thorough construction documentation. Drawings, schedules, material lists, specifications, and other types of documentation can be automatically generated by extracting detailed data from the BIM model regarding building elements, systems, and materials. By doing this, the documentation process is streamlined and the errors and inconsistencies that could arise from creating documentation by hand are reduced.

In addition, BIM reduces the possibility of errors and ensures consistency throughout documents by doing away with the necessity for duplicate or manual data entry. Because the data in the BIM model are interconnected, any modifications made to the model automatically update the related documentation, ensuring accuracy and consistency all the way through the project.

For the benefit of all project stakeholders, BIM serves as a collaborative platform that promotes coordination and communication throughout the documentation phase. The BIM model is accessible and editable by a variety of disciplines, including engineers, architects, contractors, and subcontractors, guaranteeing that the documentation reflects a common understanding of the project. This strengthens collaboration, lowers misunderstandings, and raises the standard of the documentation in total.

The ability of BIM to facilitate design validation and simulation to guarantee adherence to project specifications and quality standards is another crucial aspect of documentation improvement. Through the use of BIM model simulation, project stakeholders are able to evaluate constructability, performance, and functionality. This aids or helps in locating any possible conflicts, coordination problems, or design faults that can lower the standard of the final construction product. Through design validation, these problems are quickly and early identified and corrected, improving or enhancing the overall quality of the construction.

Construction teams may keep an eye on the actual construction process in comparison to the intended schedule and design intent with BIM's support for real-time progress tracking. Discrepancies or variations between the as-built conditions and the BIM model can be detected and utilized to initiate proactive quality control measures. Any changes to the construction schedule can be handled quickly to guarantee that quality standards and specifications are followed.

More also, the use of BIM in quality control is not limited to the construction stage. Facility management and maintenance activities can benefit from the rich data contained in the BIM model. Precise as-built records and comprehensive information regarding building elements enable effective asset tracking, lifecycle management, and maintenance planning. This raises the constructed facility's long-term performance and quality.

Building information modeling (BIM) greatly enhances quality control and construction documentation procedures. It promotes design validation and simulation, improves accuracy, consistency, and collaboration in the creation of documentation, makes collision detection and conflict resolution possible, makes real-time progress tracking easier, and helps with facility management. Construction projects can achieve better long-term performance, lower rework, better coordination, and higher quality standards by utilizing BIM.

Chapter 7: BIM and Facility Management (BIM-7D)

Building Information Modeling and Facility Management are two interconnected disciplines that play an important role in the lifecycle of a building. BIM refers to the process of creating and managing typical digital representations of physical and functional characteristics of a building or infrastructure. Facility Management, on the other hand, deals with the ongoing operations and maintenance of a built environment once it is completed. It encompasses a range of activities, including managing building systems, ensuring occupant comfort and safety, overseeing maintenance and repairs, optimizing energy usage, and managing space utilization.

The owners, operators, and occupants of buildings can gain several advantages from the integration of BIM and facility management. The following are some major benefits of using BIM in facility management:

1. Enhanced Collaboration for facility managers: For the benefit of all parties participating in the design and management stages, BIM offers a central platform. It facilitates smooth communication, coordination, and collaboration between contractors, architects, engineers, and facility managers. By integrating, miscommunications are decreased, and communication is enhanced.

2. Improved Data Management: Building Information Modelling functions or serves as an all-inclusive database for building data, containing details on the equipment, warranties, maintenance schedules, and requirements. Facility managers can use this information to prepare for equipment replacement, expedite maintenance processes (both proactive and reactive), and make well-informed decisions based on accurate and up-to-date data.

3. Efficient Space Management and interior design: Building information modeling (BIM) models can be used to maximize space use in a structure. Facility managers can evaluate occupancy trends, identify sections that aren't being used, and decide on reconfigurations or space allocations with informed decisions This optimizes space utilization and may lower expenses.

4. Predictive or Proactive Maintenance: Proactive maintenance planning is made possible by BIM models that are integrated with facility management systems. Facility managers can plan repairs or replacements, reduce downtime, and predict maintenance needs by recording and analyzing building performance data. This strategy may result in lesser costs and more operational effectiveness.

5. Sustainability and Energy Efficiency: BIM makes it easier to analyze the energy performance of buildings and identify possible energy-saving solutions. Through the integration of energy simulation tools, facility managers may maximize energy efficiency, assess the effects of design modifications, and put strategies into action to improve the sustainability of the building.

In general, a more effective and efficient management of buildings throughout their lifecycle is made possible by the integration of BIM and facility management. In the end, it gives building occupants and stakeholders a better experience by enhancing sustainability, lowering costs, streamlining operations, and improving collaboration.

7.1 Leveraging BIM for facility operation and maintenance

Managing and maintaining buildings efficiently throughout their lifecycle is made possible by leveraging BIM (Building Information Modeling) for facility operation and maintenance. This section will go into great detail about several important applications of BIM for facility operation and maintenance.

Building information modeling (BIM) functions as a centralized database that contains comprehensive details about mechanical, electrical, structural, and architectural systems. Building components, specs, and maintenance needs can be thoroughly understood by facility managers by accessing this information. They are able to locate equipment, comprehend how it works, and obtain essential documentation through it (Access to Comprehensive Building Information).

Through the integration of asset management systems, BIM can assist with proactive maintenance planning. Facility managers may immediately link BIM components to maintenance schedules, procedures, and documentation. They can more effectively organize tasks, track work orders, and create automatic maintenance plans as a result of this connectivity. Maintenance teams can obtain the necessary data for maintenance tasks, including safety procedures, equipment manuals, and historical maintenance records, by utilizing BIM as a reference (Maintenance Planning and Scheduling).

Furthermore, conflicts and clashes between various building systems can be detected with BIM models during the design and construction phase. During facility operation and maintenance, this information can be used to identify potential interferences that might impede maintenance tasks. To maintain accessibility and minimize disruptions, facility managers can visually examine conflicts in the BIM model and plan maintenance activities accordingly (Clash Detection and Conflict Resolution).

Training and virtual maintenance simulations can both benefit from the use of BIM models. Facility managers could create virtual walkthroughs of the building that replicate maintenance processes and instruct personnel on how to operate equipment and perform routine maintenance. This virtual training enhances understanding and efficiency, minimizes risks associated with receiving training on the job and toolbox talks, and increases the general effectiveness of maintenance activities (Virtual Maintenance and Training).

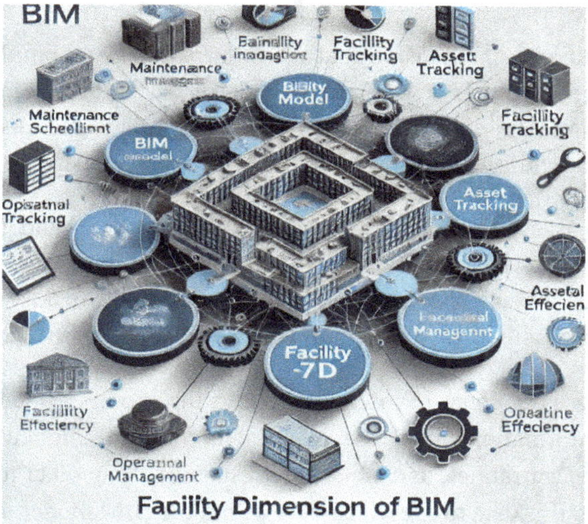

Figure 9: The facility dimension of BIM

Integrating lifecycle data with building components is made possible by BIM. Facility managers can link each component to relevant information, including expected lifespans, installation dates, and warranty details. Planning for replacements, scheduling preventive maintenance activities, and monitoring the age and performance of assets are all made easier with the use of this data. Facility managers may maximize asset performance, reduce downtime, and save maintenance costs by utilizing BIM for lifecycle data management (Lifecycle Data Management).

Building management systems, sensors, and Internet of Things devices can all be related to BIM to provide a real-time facility performance monitoring. Facility managers can identify operational efficiency and any problems relating to the building by collecting information on energy consumption, equipment health, and occupants comfort. Proactive maintenance, early fault detection, and energy-saving measures are made feasible by this data-driven approach. (Remote Monitoring and Facility Analytics).

Facility managers can increase overall operational efficiency, maximize asset performance, optimize maintenance planning, and streamline operations by leveraging BIM for facility operation and maintenance. Building information modeling (BIM) and facility management practices integrate to create a powerful toolset or toolkit for efficient building lifecycle management and maintenance.

7.2 Asset management and lifecycle integration in BIM

Incorporating comprehensive data and information about building assets at every stage of their life cycle is a key component of asset management and lifecycle integration in BIM (Building Information Modeling). This section will discuss how asset management and lifecycle integration can be achieved using BIM.

Individual assets within a building can be well identified and tagged with the use of BIM. Within the BIM model, every asset, including fixtures, systems, and equipment, can be given a special tag or identification. This makes it possible to track and manage assets easily throughout their lifecycle (Asset Identification and Tagging).

Integrating asset-specific information into the model is made easier by BIM. This contains information about the manufacturer, installation dates, maintenance records or history, warranties, and expected lifespans, among other things. Through the process of linking this data to specific assets within the BIM model, facilities managers are able to access a comprehensive asset database (Asset Information Integration).

Facilities managers are able to create proactive maintenance schedules and plans by leveraging the asset information contained in the BIM model. Maintenance activities can be effectively planned and scheduled by examining expected lifespans, maintenance requirements, and historical data of assets. This strategy maximizes asset performance while minimizing reactive maintenance (Maintenance Planning and Scheduling).

Also, predictive maintenance techniques can be implemented through BIM integration. Facility managers can get up-to-date information on the performance and health of their assets by connecting asset data with IoT sensors and condition monitoring systems. By using this data to forecast maintenance requirements, identify potential failures or issues, and plan maintenance activities appropriately, unplanned downtime may be minimized, and maintenance resources can be maximized (Predictive Maintenance).

In addition to predictive maintenance facilitation, the life cycle integration of BIM facilitates long-term strategies for asset planning and renewal. Facility managers can develop proactive asset replacement plans by evaluating asset data, such as estimated lifespans, maintenance histories, and performance patterns. This facilitates prompt replacements, lowers the possibility of equipment malfunctions or failures, and optimizes the use of budgets and resources (Replacement and Renewal Planning).

Data Handover to Facility Management: The seamless transfer of asset data from the construction phase to facility management is made possible by BIM's integration with asset management systems. The asset data in the BIM model is updated with as-built, commissioning, and maintenance record information as the building progresses through its lifecycle. Facility managers will have access to accurate and up-to-date asset information for efficient operations and maintenance because of this comprehensive data transfer or handover (Data Handover to Facility Management).

Facility managers may maximize asset performance, expedite maintenance operations or tasks, reduce downtime, and make data-driven choices throughout the lifecycle of the building by integrating asset management and lifecycle information into BIM. The moment asset data and BIM models are integrated, a strong platform for effective and efficient asset management is established.

7.3 Using BIM for renovation and retrofitting projects

In the building and construction industry, renovation and retrofitting are two vital processes that concentrate on enhancing preexisting structures. The process of restoring a structure to improve its appearance, functioning, or both is referred to as **renovation**. This frequently entails updating antiquated areas, fixing broken components, and improving the property's overall aesthetic appeal. Renovations can involve everything from minor cosmetic modifications like painting and new flooring to larger structural adjustments like adding extra rooms or doing a complete interior redesign. Conversely, **retrofitting** is the process of upgrading already-existing structures to enhance its functionality, especially with regard to sustainability, safety, and energy efficiency. Retrofitting can involve new insulation installation, energy-efficient window replacements to replace outdated ones, HVAC system upgrades, or seismic building reinforcement. Enhancing the building's usage while prolonging its lifespan and reducing its environmental impact is the primary objective of retrofitting.

Renovation and retrofitting projects can benefit greatly from the use of BIM (Building Information Modeling), which provides better planning, coordination, and execution of these intricate tasks. The table below illustrates the application of BIM to remodeling and retrofitting projects:

	BIM's assistance in remodeling and retrofitting activities	**Comprehensive process explanations**
1	Existing Conditions Assessment	Accurate and thorough documenting of the current building conditions is made possible by BIM. The existing state of the building can be captured and incorporated into the BIM model by hand modeling or 3D laser scanning. This offers a trustworthy foundation for design and renovation planning, assisting in the identification of challenges and opportunities.
2	Clash Detection and Conflict Resolution	Designers and contractors can detect any conflicts between proposed design changes and existing building elements with the help of BIM models. Prior to the beginning of construction, clashes or interferences can be found and virtually resolved by simulating the remodeling or retrofitting plans in the BIM model. This guarantees a more effective building process and reduces the need for costly rework.
3	Design Visualization and Communication	Stakeholders can more easily comprehend the suggested remodeling or retrofitting plans because of BIM's 3D visualization capabilities. Architects, engineers, clients, and other relevant stakeholders can visualize design modifications, explore alternative solutions, and offer comments with the help of the BIM model. As a result, project participants collaborate better and communicate more effectively, which helps to make more informed decisions and clarify the project's objectives
4	Design Iteration and Analysis	BIM facilitates iterative design processes, where multiple design options can be examined and explored. Design changes and options can be swiftly integrated in the BIM model, allowing for analysis of their impact on the building's performance, energy efficiency, structural integrity, and more. This iterative design process makes informed decisions easier and ensures that the remodeling or retrofitting plans are in line with the objectives and specifications of the project.

5	Cost Estimation and Material Planning	Accurate quantity takeoffs and cost analysis are made possible by the integration of BIM models with material management and cost estimation tools. Through the integration of cost information with BIM components, project stakeholders can provide comprehensive cost estimates, streamline the acquisition of materials, and evaluate the project's financial viability for renovation or retrofitting.
6	Phased Construction and Sequencing	Phased construction is frequently necessary for renovation and retrofitting projects to reduce disruptions and maintain building functionality. Phased construction sequence planning and visualization are made easier by leveraging BIM. Project teams can optimize sequencing, spot possible conflicts or coordination challenges, and ensure a seamless transition from one phase to the next by simulating each one in the BIM model.
7	As-Built Documentation	As-built information can be captured through BIM during renovation or retrofitting process of a building. Facility managers can receive an accurate and up-to-date digital representation of the renovated or retrofitted building by updating the BIM model with modifications made during construction. For prospective future renovation projects, facility management, and maintenance, this as-built documentation becomes a valuable resource.

Table 8: BIM's assistance in remodeling and retrofitting activities (Source: author)

Stakeholders may increase communication and collaboration, expedite the planning and execution processes, lower risks and errors, maximize cost and material management, and provide better project outcomes by utilizing BIM for remodeling and retrofitting projects. These projects benefit from the usage of BIM, which increases productivity, accuracy, and project success overall.

Chapter 8: BIM and Sustainability (BIM-6D)

Sustainability and Building Information Modeling (BIM) are closely connected in the architecture, engineering, and construction (AEC) industries. A building's or infrastructure's functional and physical characteristics are represented digitally using BIM. Before commencement of construction, it enables stakeholders to collaborate, evaluate or analyze, and simulate designs.

The concept of sustainability, on the other hand, refers to the process of creating and maintaining structures that support social and economic well-being while having little to no negative effects on the environment. There are several benefits when BIM and sustainability principles are integrated.

Energy performance can be optimized during the design stage with the use of BIM. Architects and engineers may simulate and assess a building's energy usage, pinpoint possible areas for improvement, and make well-informed decisions to improve energy efficiency with the use of energy analysis tools that integrate with BIM software (Energy Efficiency).

Also, the analysis and comparison of various building or construction materials and their effects on the environment is facilitated by BIM. Through the consideration of many factors including embodied energy, life cycle assessments, and

recycled content, designers can make well-informed decisions regarding materials that contribute to sustainability and reduce carbon emissions (Material Selection).

Accurate quantity takeoffs and exact project planning are made possible by BIM, and these capabilities can reduce the generation of waste during construction. Building Information Modeling (BIM) minimizes overordering and excessive waste on construction sites by optimizing material quantities and fostering better coordination among project stakeholders (Waste Reduction).

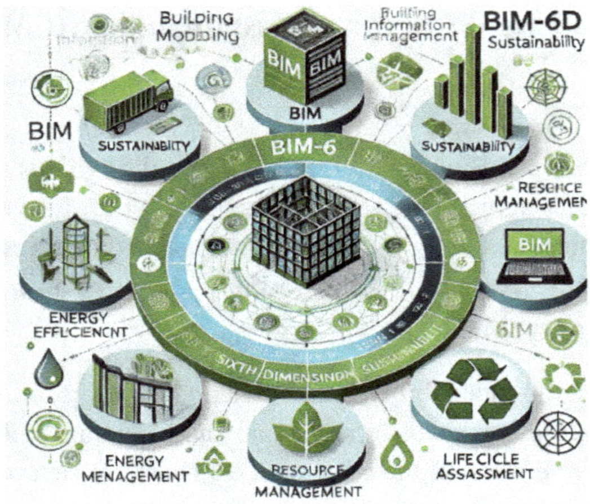

Figure 10: Showing BIM and Sustainability BIM-6D

In addition, BIM can be utilized throughout a building's operation and maintenance phases as well as its whole lifecycle. It enables facility managers access to important and useful information regarding equipment, systems, and maintenance schedules, which promotes effective building operations, lower energy consumption, and lower life-cycle costs. This is referred to as lifecycle analysis.

More significantly, BIM provides sophisticated simulation and visualization tools that let stakeholders and designers see how their choices will affect the environment. They are able to assess factors like wind flow, daylighting, and solar shading through simulations, which helps them design sustainable buildings and optimize building performance.

The data-rich models provided by BIM can be used for continuous facility management, enabling effective upkeep and operation. Facility managers may ensure long-term sustainability by tracking energy use, keeping an eye on equipment performance, and identifying improvement opportunities with the integration of sustainability-related data.

The architecture, engineering, and construction industry can make great strides in developing constructed environments that are green, energy-efficient, and socially conscious by fusing BIM approaches with sustainability concepts. Throughout a building's lifecycle, this integration promotes better decision-making, lower environmental impacts, increased resource efficiency, and long-term sustainability.

8.1 Incorporating sustainable design principles in BIM

Building or infrastructure projects are ideal candidates to incorporate sustainability during the design phase of a project. To choose building materials, equipment, and fixtures that meet BREEAM and LEED certification requirements while being environmentally friendly, architects, engineers, and other construction professionals can collaborate during the design stage. Integrating environmentally friendly or ecologically practices into the digital design and construction process is the process of incorporating sustainable design principles into BIM (Building Information Modeling). A few significant considerations to take into consideration while incorporating sustainability into BIM will be discussed in the table section below.

	Design Stage key Consideration and Practices while Incorporating Sustainability into BIM in Projects	**Discussion**
1	Energy Efficiency Analysis	During the design process, engineers and architects can simulate and analyze energy performance with the use of BIM software. To optimize energy efficiency and reduce carbon or greenhouse gas emissions, designers can assess factors such as building orientation, solar or sun exposure, and insulation by using energy analysis tools.
2	Material Selection and Life Cycle Assessment	Designers can evaluate and compare different building materials based on the material impact on the environment with the aid of BIM. By incorporating life cycle assessment (LCA) data into BIM models, material selection activity will be carried out effectively by project professionals with informed decision, considering things like end-of-life considerations, recycled content, and embodied energy.
3	Water Efficiency	Using BIM to analyze systems like rainwater harvesting, graywater recycling, and energy-efficient plumbing fixtures can assist optimize water usage. Water flow and consumption patterns can be simulated by designers and construction professionals in order to find ways to save water and reduce waste.

4	Daylighting and Natural Ventilation	Accurate analysis of natural ventilation and daylighting strategies is made possible by BIM. Designers can maximize natural light and fresh air by strategically placing windows, shading devices, and ventilation openings by simulating the sun's path and airflow patterns. This helps to reduce the need for artificial lighting and mechanical cooling.
5	Renewable Energy Integration	The integration or incorporation of renewable energy systems into architectural or building designs can be made easier by the use of BIM. To lessen dependency on fossil fuels, designers and construction professionals can choose the best locations and sizes for solar panels, wind turbines, and other renewable energy technologies or strategies by analyzing solar potential and wind patterns.
6	Waste Management and Recycling	Construction site waste management practices can be optimized with the aid of BIM. It helps or contributes to the reduction of waste generation by accurately estimating or costing material quantities needed and scheduling or coordinating deliveries. Additionally, BIM can also help with the planning of effective or efficient waste diversion and recycling strategies.
7	Indoor Environmental Quality	Indoor environmental factors such as acoustics, air quality, and thermal comfort can be simulated and analyzed using BIM. Designers can enhance occupant comfort and well-being by creating healthier indoor environments through the optimization of HVAC systems, ventilation rates, and material options.
8	Green Building Certifications	Building Research Establishment Environmental Assessment Method, commonly known as BREEAM and LEED (Leadership in Energy and Environmental Design), are two examples of green building standards that BIM may help with the documentation and certification process. The data and documentation needed to obtain certifications for sustainable buildings can be found in BIM models.

Table 9: Design Stage key Consideration and Practices while Incorporating Sustainability into BIM workflow

Construction professionals such as engineers, architects, and construction managers can create more ecologically or environmentally friendly and energy-efficient buildings by incorporating sustainable design concepts into the BIM workflow. Since BIM is collaborative and information-rich, it facilitates or enhances data-driven decision-making, building or structural performance optimization, and the development of sustainable built environments.

8.2 Energy analysis and simulation in BIM

Building information modeling (BIM) is an important instrument that helps architects, engineers, and designers assess and enhance a building's energy performance. Two fundamental characteristics of BIM are energy analysis and simulation. This section will explain how simulation and energy analysis function inside the BIM framework.

1. Building Geometry and Data: The first step in building information modeling (BIM) is to digitally depict the geometry of the building, including the walls, roof, floors, and openings like windows and doors. These elements can be precisely modeled using BIM software, which captures their dimensions, orientations, and thermal properties.

2. Energy Analysis Tools: Energy analysis tools are integrated into BIM software, which uses building geometry and data to simulate energy performance and consumption. These software programs calculate variables, including heat transfer, solar radiation, and thermal loads using algorithms and the basic principles of building physics.

3. Thermal Analysis: Thermal analysis, which evaluates the heat flow within a building and its components, is made feasible using BIM. Designers can optimize insulation, fenestration, and shading strategies by identifying potential areas of heat gain or loss through the use of heat transfer simulations through walls, roofs, and windows.

4. Daylighting Analysis: The distribution of natural light within a building can be simulated and analyzed with BIM. Designers can determine whether artificial lighting is necessary and evaluate daylight availability by taking into account the building's orientation, location, and window placement or layout. With this information, one may enhance natural lighting while consuming the least amount of energy by optimizing window sizes, glazing types, and shading devices.

5. HVAC System Analysis: HVAC (heating, ventilation, and air conditioning) system analysis is made feasible by BIM. Designers can evaluate the performance of HVAC systems in terms of energy efficiency or economy,

airflow distribution, and thermal comfort by simulating their behavior within the building. This analysis helps in ductwork layout optimization, HVAC equipment selection, control strategy development for efficient operation.

6. Energy Consumption Evaluation: Energy consumption can be evaluated using BIM through simulated building performance. By the consideration of many factors such as occupancy, equipment usage, and climate data, designers can determine the energy demands of the building and identify areas where energy efficiency and reduction can be achieved.

7. Renewable Energy Integration: The integration of renewable or clean energy systems or mechanisms, like solar panels or wind turbines, can also be analyzed or assessed through BIM. By the process of simulating these systems' performance and output based on the location and characteristics of the building, designers are able to evaluate the feasibility and potential advantages or benefits of incorporating renewable or clean energy sources to the building's energy mix.

8. Optimization and Iteration: Design processes can be iterated because of the interactive nature of BIM. The building's geometry, systems, or materials can all be changed. To evaluate the impact of these adjustments, designers might run or perform energy simulations. Better energy performance, energy-saving potential, and design refinement are all made possible by this iterative process.

Through the application of energy analysis and simulation strategy in the BIM environment, designers may make well-informed decisions and optimize or enhance energy-saving solutions by gaining useful insights into a building's energy performance. Through the use of these tools or technologies, project stakeholders can produce or create designs that are more environmentally or ecologically friendly and energy-efficient, hence reducing the building's lifecycle operating costs.

8.3 Evaluating the environmental impact of buildings through BIM

Building Information Modeling (BIM) is a useful tool for assessing how buildings affect the environment throughout their lifecycle. Embodied Energy and Carbon Assessment, Construction and Operational Energy Analysis, Water Usage and Efficiency, Waste Generation and Management, Environmental Performance Metrics,

Green Building Certification, Sensitivity Analysis and Scenario Testing are some important components of using BIM to evaluate and mitigate the environmental effects.

Building Information Modeling (BIM) makes it possible to evaluate a building's embodied energy or the energy consumed in the extraction, production, transportation, and assembly of building materials. Designers are able to measure the carbon footprint of building materials and make well-informed decisions to minimize embodied energy and carbon emissions by incorporating life cycle assessment (LCA) data into BIM models (Embodied Energy and Carbon Assessment)

The energy consumption of a building during its operational phase can be simulated and analyzed using BIM. Energy-saving opportunities can be identified, and energy demands can be estimated by designers by taking into account different factors like HVAC systems, lighting, appliances, and occupancy patterns. The selection of energy-efficient technologies and strategies to lower operational energy use is aided by this analysis (Construction and Operational Energy Analysis).

The efficiency and use of water in a building can also be evaluated through BIM. Designers can evaluate water demand and identify areas that require water conservation by simulating plumbing systems, water flow, and consumption patterns. In order to reduce water usage and its impact on the environment, BIM makes it easier to include water-saving systems like rainwater harvesting, graywater recycling, and energy-efficient plumbing fixtures (Water Usage and Efficiency)

More also, waste generated during construction and operation can be evaluated and managed with the use of BIM. Designers can reduce waste and the environmental impact of construction activities by precisely calculating material quantities and streamlining construction processes. Planning for efficient waste management strategies, such as recycling and waste diversion, can also be facilitated by BIM (Waste Generation and Management).

Building environmental performance metrics, such as carbon emissions, energy use, water use, and waste generation, can be calculated and presented using BIM. These indicators give stakeholders important information about a building's overall sustainability and its environmental impact, enabling them to monitor and track the building's progress toward sustainability goals (Environmental Performance Metrics).

The documentation and analysis needed for green building certification processes like BREEAM, fully known as Building Research Establishment Environmental Assessment Method) and LEED (Leadership in Energy and

Environmental Design) can be supported by BIM. BIM models facilitate the certification process by offering the information and documents required to prove compliance to sustainability standards (Green Building Certification).

Sensitivity analysis and scenario testing to evaluate the environmental impact of design alternatives are made possible by BIM's simulation capabilities. In order to optimize the sustainability performance of the building and choose the most ecologically friendly or environmentally favorable options, designers can assess the effects of various material decisions or choices, energy-efficient systems, or integration of renewable energy sources (Sensitivity Analysis and Scenario Testing).

Architects, engineers, construction managers, and other relevant stakeholders may assess the environmental impact of buildings from the start of construction to the stage of operation by utilizing the power of BIM. The analytical and data-rich capabilities of BIM enable sustainable design strategies, help make well-informed decisions and help create resource- and environmentally conscious buildings.

Chapter 9: BIM Safety Model (BIM-8D)

The integration of safety-related information and analysis within the wider framework of a Building Information Modeling (BIM) system is referred to as a BIM Safety Model. With the use of BIM technology, building project digital representations can be created, managed, and shared, enabling stakeholder collaboration and well-informed decision-making throughout the construction process.

The process of integrating safety into a BIM model involves including specific safety-related data, like emergency response plans, risk assessments, hazard identification, and safety regulations or legislation. Project teams can see and assess any potential safety risks, develop proactive safety measures, and simulate safety scenarios by incorporating this information into the BIM model.

The BIM Safety Model improves coordination and communication between contractors, engineers, architects, and other construction project professionals. It gives stakeholders a comprehensive and visual representation of the project, enabling them to recognize and address safety concerns early on. This proactive strategy reduces accidents, mitigates risks, and enhances overall project safety.

Some key features of a BIM Safety Model may include:

1) Hazard identification and risk assessment: Identifying potential hazards and assessing or evaluating their associated risks within the project model.

2) Safety regulations and compliance: Integrating pertinent safety standards, regulations, and codes into the BIM model to guarantee compliance throughout the project lifecycle.

3) Clash detection: Identifying or spotting clashes or conflicts between safety-related components, elements, or systems within the model, like structural elements obstructing fire exits.

4) Simulation and visualization: Visualizing and creating evacuation routes, simulating safety scenarios, and testing emergency response plans to assess their effectiveness may all be done with the help of the BIM model.

5) Safety documentation: Generating or producing safety-related paperwork or documentation directly from the BIM model, such as safety training materials, inspection reports, and plans.

6) Maintenance and facility management: incorporating safety-related data into the BIM model for future renovations, maintenance, and ongoing facility management.

Construction project teams can improve overall safety outcomes, minimize potential hazards, and improve safety planning by using the BIM Safety Model. The safety focused data integration and the collaborative nature of BIM technology, helps foster a safer and more efficient construction process.

Chapter 10: Lean construction facilitated by BIM (BIM-9D)

The philosophy and approach of lean construction involves streamlining the construction process in order to reduce or eliminate waste, boost production and complete projects more quickly and efficiently. Technology such as Building Information Modeling (BIM) plays a crucial role in facilitating lean construction practices. This chapter will go into great detail on how BIM helps and facilitates lean construction.

Architects, builders or construction managers, engineers, contractors, and subcontractors are among the project stakeholders who may collaborate and exchange information easily through BIM. This collaborative platform lowers barriers to communication, makes coordination easier, and for real-time and quick decision-making. Teams that collaborate more effectively are able to identify or spot potential issues early on and find better and lasting solutions, which reduces delays and rework (Enhanced collaboration).

BIM makes it possible to create precise and comprehensive 3D models that incorporate structural, architectural, and MEP (mechanical, electrical, and plumbing) systems. With the help of these models, teams may virtually see and evaluate multiple design options for the project. Teams can enhance construction efficiency, minimize material waste, and optimize layouts by simulating and evaluating design alternatives within the BIM environment (Design optimization).

Conflicts or clashes amongst different building elements or systems, such as those between ductwork and structural components or elements, can be automatically detected by the model with the use of BIM. By identifying conflicts early on, teams can minimize the need for costly modifications and rework before construction even commences (Clash detection and coordination).

In addition, off-site construction techniques and prefabrication can be integrated with the support of BIM. Teams are able to effectively plan and coordinate the off-site manufacturing of building components by utilizing the comprehensive information contained in the BIM model. Prefabrication minimizes waste, enhances quality control, and lowers the number of craftsmanship and on-site labor (Prefabrication and off-site construction).

The development of detailed constructing schedules based on the 3D model is made feasible using BIM. Teams may enhance construction flow, identify any potential bottlenecks, and optimize the schedule by integrating time-related information and visualizing the construction sequence. This makes it possible to allocate resources more effectively, minimize downtime, and increase project efficiency (Construction sequencing and scheduling).

Also, Budget control and accurate cost estimation are made possible by the useful information included in BIM models. Teams can generate accurate cost estimates or reliable cost forecasts and keep better track of project expenditures by connecting the model to cost databases and estimating software. This makes it possible for improved financial planning and management, which promotes lean construction methodologies or practices (Cost estimation and budget control).

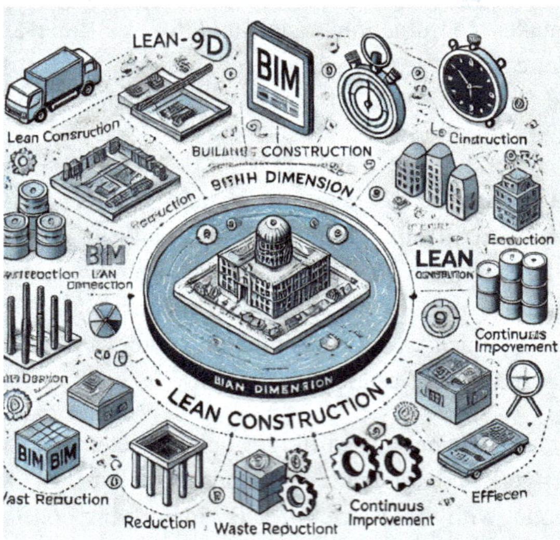

Figure 11: BIM facilitating Lean Construction principles

All things considered, BIM technology offers the instruments and capacities required to optimize designs, reduce waste, enhance collaboration, and streamline construction processes. Projects can achieve improved quality, lower costs, faster project delivery, and higher productivity by integrating lean construction principles with BIM.

Chapter 11: BIM and Industrialized Construction (BIM-10D)

Industrialized construction methods rely heavily on the support and facilitation of Building Information Modeling (BIM). Industrialized construction is the practice of prefabricating building components off-site using standardized methods and advanced manufacturing techniques. This reduces construction time, increases efficiency, and improves quality control.

Prefabricated buildings are made off-site, and the building materials are subsequently transported to the location for assembly. Industrial construction has a number of benefits, some of which are reduced labor requirements on the construction site, flawlessly built or fabricated building components at standard temperature conditions, and minimal to nonexistent weather effects during fabrication.

11.1 BIM complements and enhancement of industrialized construction

In numerous ways, BIM enhances and complements industrialized construction. BIM makes it possible to create intricate 3D models that accurately depict the systems and components of a building. With the help of these models, prefabricated element designs may be visualized and coordinated, ensuring a perfect or seamless fit and eliminating clashes or conflict. BIM aids in the design's manufacturing optimization, enabling fabrication processes to be carried out with accuracy and precision (Design coordination and visualization).

Additionally, prefabrication process planning, and optimization are aided by BIM software. The best and most economical or cost-effective fabrication techniques, material quantities, and assembly sequences can be found using the digital models. Accurate shop drawings and specifications can also be generated with the help of BIM, assuring uniformity or consistency and reducing fabrication process errors (Prefabrication planning and optimization)

The logistics of off-site fabrication can be planned and coordinated with the help of BIM. The BIM model's comprehensive information aids in determining the fabrication facility's material handling, assembly order or sequence, and transportation requirements. In addition, supply chain management, tracking inventory, and coordinating the transportation or delivery of prefabricated components to the construction site can all be facilitated by BIM (Off-site assembly and logistics).

Also, when it comes to on-site assembly and installation, BIM models offer invaluable information. Construction workers can precisely position and connect prefabricated components, reducing errors and improving productivity by integrating the BIM model with augmented reality or other visualization tools or techniques. The progress of installation activities can also be tracked and managed with the use of BIM (On-site installation and assembly).

Furthermore, industrialized construction's continuous maintenance and facility management are supported by BIM's robust data integration capabilities. Prefabricated component specifications, maintenance schedules, and performance data can all be found in the BIM model. This information can be used to ensure the long-term sustainability and efficiency of the constructed facility through asset management, lifecycle analysis, and preventive maintenance planning (Maintenance and facility management).

Project teams can derive numerous benefits by integrating BIM technology with industrialized construction methods, including:

- Increased or Improved productivity and shortened construction schedules.
- Enhanced or better-quality control or assurance and reduced rework.
- Improved safety and security through less on-site activities.
- Cost savings through waste reduction and optimal material use.
- Improved collaboration, coordination, and communication between project teams or stakeholders.
- Effective and Efficient facility maintenance and management throughout the building's lifecycle.

As a whole, BIM is an effective tool that facilitates the integration and coordination of industrialized construction processes, leading to more efficient workflows, increased output, and higher-quality construction outputs.

Chapter 12: BIM Standards and Regulations

Building information modeling, or BIM, standards and regulations are a set of guidelines and requirements designed to foster the effective implementation and use of BIM technology in the construction sector. These regulations and standards serve to ensure consistency, interoperability, and collaboration among every project stakeholder involved in the design, construction, and operation of infrastructure or building projects and buildings.

12.1 Some key aspects of BIM standards and regulations

International guidelines such as ISO 19650, which offer a global framework for managing information over the whole lifecycle of a built asset, are important aspects of BIM standards. National regulations, customized for specific countries, make sure that BIM practices adhere to local legislation and industry requirements. Furthermore, industry-specific guidelines provide comprehensive instruction on the implementation of BIM in particular sectors, ensuring that the different needs of various construct disciplines are met.

Figure 12: BIM national and international guidelines

For BIM adoption to be effective, it is essential to comprehend and abide by these standards and regulations since they support the maintenance of high-quality results, reduce risks, and foster better project collaboration. Some of the most significant aspects of BIM regulations and standards are:

1. **Information Exchange Formats**: Standardized file formats and data structures are defined by BIM standards to ease the exchange of data between different software applications. The Industry Foundation Classes (IFC) format is widely utilized for enabling interoperability between diverse BIM software platforms.

2. **Level of Development (LOD):** LOD definitions, which specify the degree of accuracy and level of detail of the information in a BIM model at different project stages, are often included in BIM standards. Project teams can better understand the anticipated level of information and ensure consistency in BIM deliverables with the use of these definitions.

3. **Data and Model Naming Conventions**: Within a BIM project, standards offer guidelines for uniformed or consistent naming conventions for data and model elements. This guarantees information sharing between project stakeholders is consistent, done with ease, and clear.

4. **Classification Systems**: Building elements and components can be organized and categorized using classification systems like UniFormat, OmniClass, or MasterFormat, which are integrated into BIM standards. These systems make it possible to consistently classify and group information within BIM models, which makes it easier to retrieve and analyze information.

5. **BIM Execution Plans (BEP):** The BIM responsibilities, processes, and requirements that are unique to a project are outlined in a BEP. It may include information about the BIM standards to be followed, the intended level of detail, information exchange protocols, and project participant coordination procedures.

6. **Government Mandates and Regulations**: Some countries or regions have passed regulations mandating the use of BIM in projects involving public infrastructure. These regulations define certain BIM standards and requirements that must be adhered to in order to ensure data exchange, collaboration, and consistency throughout the lifecycle of the project.

7. **Quality Control and Validation**: Guidelines for BIM model validation and quality control are often included in BIM standards. This ensures the information in the models is accurate, consistent, and reliable, which reduces errors and makes it easier to make wise decisions during the construction and operation stages.

It's important to keep in mind that BIM regulations and standards might differ between countries, regions, and even individual projects. To ensure compliance and interoperability, it is essential that you review the relevant standards and regulations that are unique to the jurisdiction or project you are working on.

12.2 Overview of international BIM standards and guidelines

Several international BIM guidelines and standards have been developed to facilitate the consistent implementation and compatibility or interoperability of BIM technology among different countries and regions.

Building information modeling can be adopted and implemented globally with the help of international standards and guidelines. Some of these standards, like ISO 19650, outline and describe best practices for managing information during the whole construction project lifecycle, from initial design to demolition. By encouraging a standard approach, these guidelines make it possible to collaborate, exchange data, and manage projects consistently across different regions and industries. Following these international standards makes sure that projects meet or comply with quality, safety, and sustainability benchmarks, which increases efficiency and reduces the risk of errors in complex construction projects.

12.2.1 An overview of some prominent international BIM standards and guidelines

1. **ISO 19650**: An international standard known as ISO 19650 is used to manage information using BIM across a built asset's whole life cycle. It offers a structure or framework for information management, organization, and exchange in a collaborative BIM environment. There are two parts to ISO 19650: While Part 2 offers guidance on how to use BIM, Part 1 focuses on concepts and principles.

2. **IFC (Industry Foundation Classes):** The buildingSMART organization developed the open and neutral file format known as IFC. It facilitates

interoperability between different disciplines and project stages as well as the exchange of BIM data between various software programs or applications. The international BIM community has embraced IFC as a common data exchange format.

3. **COBie (Construction Operations Building Information Exchange):** During the handover of a construction project, facility information can be sent using the non-proprietary COBie data standard. It offers a structured approach for capturing and delivering asset data, including equipment lists, warranties, maintenance schedules, and spare parts, that is necessary for facility management.

4. **PAS 1192 series**: The British Standards Institution (BSI) developed the PAS 1192 series of standards, which provide a comprehensive basis for implementing BIM in the UK. PAS 1192-2 (Specification for information management for the capital/delivery phase of construction projects), PAS 1192-3 (Specification for information management for the operational phase of assets using BIM), and other documents are all included in the series.

5. **AEC (Architecture, Engineering, and Construction) BIM Protocol:** The UK government established the AEC BIM Protocol as a framework to define contractual and legal obligations for BIM implementation. It lays out the processes for information exchange and collaboration throughout a project's lifecycle as well as the rights, responsibilities, and liabilities of project participants.

6. **GSA (General Services Administration) BIM Guide Series**: The U.S. General Services Administration developed the GSA BIM Guide Series, which offers guidelines and best practices for using BIM in the public sector. It covers a variety of aspects, such as model development, quality control, project planning, and BIM execution planning.

7. **Singapore BIM Guide**: The Building and Construction Authority (BCA) of Singapore published the Singapore BIM Guide, which lays out the standards or guidelines and best practices or requirements for BIM in public sector projects. It addresses subjects like data exchange standards, BIM collaboration processes, LOD definitions, and deliverables for BIM.

These are but a handful of the international BIM guidelines and standards. It's crucial to keep in mind that different countries and regions might have their own distinct regulations and standards that are suitable to their specific needs. To ensure efficient BIM implementation and information exchange whether working on international projects or collaborating with stakeholders from different regions, it's critical to understand and comply by the relevant standards and guidelines.

12.3 Compliance requirements and implementation strategies

Depending on the specific regulations, standards, and project requirements or objectives, there may be differences in the BIM compliance requirements and implementation strategies. However, these are some general factors to keep in mind while trying to comply with regulations and effectively use BIM.

1. Familiarize Yourself with Applicable Standards: To start with, make sure you are familiar with the BIM standards, regulations, and guidelines that are pertinent to your project or area of jurisdiction. This comprises studying documents like ISO 19650, local BIM mandates, and industry-specific standards. Identify the precise requirements or specifications for data formats, LOD definitions, information exchange, and other relevant matters.

2. Develop a BIM Execution Plan (Point cloud): The implementation of BIM throughout the project will be outlined in a BIM Execution Plan. It should cover the requirements specific to the project, team member roles and responsibilities, information exchange protocols, LOD expectations, and coordination strategies. The BEP should guide all project participants and be in line with compliance requirements.

3. Establish Information Exchange Requirements: Clearly state what data formats, file naming conventions, and coordination procedures are required for information exchange. At each project stage, identify specific deliverables and the necessary level of detail. This will ensure uniformity or consistency and interoperability amongst the project stakeholders.

4. Implement a Common Data Environment (CDE): A central platform for managing, sharing, and storing project data is called a CDE. It makes collaboration easier and ensures that every construction professional involved in the project has access to the most recent BIM models and related documentation. By providing a controlled environment for version control and data exchanges, a CDE's implementation helps in enforcing compliance.

5. Collaborate and Coordinate with Stakeholders: For a successful BIM implementation, project stakeholders must effectively collaborate and coordinate. It is important to hold regular meetings, workshops, and coordination sessions to address any issues, resolve conflicts, and make sure that every stakeholder or everyone is making an effort to comply with the requirements.

6. Use BIM Authoring and Analysis Tools: Make use of BIM authoring tools that comply with compliance requirements by supporting the necessary data formats and interoperability standards. Efficient modeling, information management, clash detection, and analytical capabilities ought to be made possible by these tools. Choose software systems that can enable seamless exchange of data and are compatible with the required file formats, such as IFC.

7. Establish Quality Control Procedures: To ensure or assure the accuracy and reliability or dependability of BIM models and data, set quality control measures in place. This includes validation against compliance conditions, detection of clashes, and routine model checks. Establish, define, and enforce standards for data integrity, naming conventions, and model development. Perform or conduct regular audits to evaluate compliance of the defined standards.

8. Provide Training and Support: Invest in training courses to enhance project shareholders BIM knowledge. To assist or support team members in adopting and successful implementing BIM processes, provide technical assistance and guidance. This guarantees that all stakeholders involved in the project have the necessary skills, knowledge, and abilities needed to comply with requirements and efficiently use BIM tools.

9. Monitor and Review Implementation: Review compliance with established standards and regulations and track the implementation of BIM on a regular basis. As necessary, update the implementation strategy, address any non-compliance issues, and identify areas for improvement. Throughout the course of a project, continuous tracking and evaluation help in maintaining compliance.

Keep in mind that based on the specific project at hand, the jurisdiction, and the industry, there may be differences in the implementation strategies and compliance requirements. It is crucial or important to adapt these strategies to fit your specific context's needs and comply to all applicable or relevant standards and regulations.

12.4 Interoperability challenges and industry initiatives

In the context of BIM, interoperability challenges are the obstacles or difficulties that arise when integrating and exchanging information between various software applications, platforms, and stakeholders. These obstacles can impede data sharing or exchange, collaboration, and the smooth or seamless flow of information within the construction industry. In order to tackle these obstacles, a number of industry initiatives have been established. The section will discuss some typical interoperability challenges and the corresponding industry initiatives.

Different software programs use proprietary file formats, which make them not easily compatible with other platforms. The seamless exchange of BIM data may be hampered by this. Industry initiatives like COBie and the Industry Foundation Classes (IFC) have been developed to address this difficulty. While COBie provides a standardized format or framework for conveying facility information, IFC is an open standard file format that encourages interoperability (Proprietary File Formats).

Effective information integration and analysis is hampered by the potential usage of varying data structures, naming conventions, and classifications in BIM models from various stakeholders. Industry initiatives like Open Geospatial Consortium (OGC) and buildingSMART International are working to develop and promote common data standards and systems of classification in order to address this. The objective of these initiatives is to create uniform or consistent classification schemes and data structures to enhance interoperability (Inconsistent Data Structures).

Also, stakeholders may have different LOD definitions, which could result in inconsistencies in the accuracy and level of detail in BIM models. The LOD framework outlines the information needed at different stages of the project. An industry initiative called the BIMForum LOD Specification offers guidelines for defining and describing or communicating the level of development in BIM models, hence fostering better interoperability (Level of Development (LOD) Definitions).

The inability of different software tools or programs used for various tasks in the construction process to seamlessly integrate with one another frequently impedes collaboration among project stakeholders. Open standards and software solutions that are compatible with one another are encouraged by industry initiatives such as the Open BIM movement. Open BIM encourages the use of software programs or tools that facilitate data interoperability and better collaboration by allowing data to be exchanged through defined or standardized formats (Integration of Software Tools).

More also, the efficient transfer or flow of information across different software applications might be hampered by inconsistent data exchange protocols and communication standards. The goal of industry initiatives like Application Programming Interfaces (APIs) and the Common Data Environment (CDE) is to standardize data exchange protocols and facilitate smooth or seamless integration between various software platforms (Data Exchange Protocols).

And uncertainties and issues with compatibility between software applications may arise from the lack of standardized interoperability testing and certification processes. In order to address this, initiatives such as the National BIM Standard-United States (NBIMS-US) offer guidelines for certification and testing processes, guaranteeing that software applications meet or adhere to specified interoperability requirements (Lack of Interoperability Testing and Certification).

Diverse industry alliances and collaborative efforts have surfaced to address interoperability issues collectively. To improve interoperability in the construction industry, groups like buildingSMART International, the Open Geospatial Consortium (OGC), and the Construction Industry Institute (CII) strive to develop standards, promote best practices, and foster collaboration between stakeholders (Industry Alliances and Collaborative Efforts).

To address interoperability issues, these industry initiatives seek to develop best practices, encourage collaboration, and promote open standards. The construction industry can achieve increased efficiency and productivity throughout the project lifecycle by implementing these standards and initiatives, which will additionally foster data exchange and collaboration.

Chapter 13: The Future of BIM

A digital process known as BIM (Building Information Modeling) makes it feasible to create and manage information on a building or infrastructure project at any point in its lifecycle. Future innovations and trends are anticipated to influence the future course of BIM in several of ways.

Firstly, it is anticipated that BIM will continue to be widely adopted in the architecture, engineering, and construction (AEC) sector. As more countries realize the advantages of BIM in terms of efficiency, collaboration, cost savings, and sustainability, this increased adoption will result in further standardization and integration of BIM processes (Increased Adoption).

Cloud computing or technology will be used in BIM workflows even more frequently. Project stakeholders will be able to access and collaborate on BIM models in real-time from any location with cloud-based collaboration platforms, which will enhance collaboration and streamline communication (Cloud-Based Collaboration).

Furthermore, AI and ML technologies will be very important to BIM in the future. These tools or technologies make it possible to automate tasks like energy analysis, material optimization, and clash detection, which facilitates quicker and more accurate decision-making. In order to provide insights for better project planning and design, AI-powered algorithms can also analyze historical project data (Artificial Intelligence and Machine Learning).

Sensor technology and the Internet of Things (IoT) will be increasingly closely integrated with BIM. BIM models will be able to incorporate real-time data from sensors integrated into infrastructure and buildings, enabling enhanced asset performance analysis, predictive maintenance, and monitoring throughout their entire lifecycle (Integration with IoT and Sensor Technology).

The manner in which BIM models are visualized or experienced will change as virtual reality (VR) and augmented reality (AR) technologies advance. Better design understanding and decision-making can be facilitated by VR and AR's immersive and interactive experiences, which let stakeholders interact with BIM models more seamlessly with great ease (Virtual and Augmented Reality).

Beyond the phases of design and construction, BIM encompasses facilities management and maintenance for the entire lifecycle of a building or infrastructure. As a central repository for asset data, BIM models will facilitate effective operations, maintenance, and refurbishment or renovation projects activities (Lifecycle Management and Facilities Maintenance).

Additionally, BIM will keep advancing sustainability goals by integrating simulation and energy analysis tools. With the use of BIM models, designers will be able to maximize energy efficiency, analyze the influence of design decisions on sustainability, and assess the environmental performance of infrastructure and buildings (Sustainability and Energy Efficiency).

All things considered, BIM has enormous potential for enhancing collaboration, delivering better project outcomes, and maximizing the lifecycle of built assets. BIM will be essential to changing the AEC sector and allowing smarter, more sustainable built environments as technology evolves.

13.1 Emerging trends and advancements in BIM technology

In BIM technology, there are advancements and trends. Most of these technologies are easily integrated into BIM and carry out tremendous tasks in construction projects, whether they are new construction, renovation, or retrofit projects. These technologies improve project performance when integrated into BIM. Of course! The latest developments and trends in BIM (Building Information Modeling) technology are listed below.

1. Cloud-Based BIM: Workflows in BIM are being revolutionized by cloud technologies. Anywhere in the world, project stakeholders may collaborate and share data in real time through cloud-based BIM platforms. This accessibility facilitates project management and enhances coordination.

2. Mobile BIM: The development of BIM applications customized for smartphones and tablets is a result of the widespread use of mobile devices. Field collaboration, issue tracking, and progress monitoring are made easier with mobile BIM's on-site access to models, drawings, and project information.

3. 4D and 5D BIM: By incorporating time into 3D models, 4D BIM produces a visual representation of the construction schedule. Better project scheduling, planning, and resource allocation are made possible by this. More cost and resource data integration are made possible by 5D BIM, which makes precise cost estimation, quantity takeoffs, and project budgeting feasible.

4. Reality Capture and Laser Scanning: Laser scanning and photogrammetry technologies are two examples of reality capture technologies that make it possible or feasible to create or develop highly accurate 3D models from actual environments or structures. These models facilitate clash detection, as-built documentation, and renovation projects when integrated into BIM workflows.

5. Generative Design: Algorithms are used in generative design to explore multiple design options and optimize performance according to specified criteria. Generative design algorithms can produce innovative and efficient design solutions by utilizing BIM data, which can save time and enhance project outcomes.

6. AI and Machine Learning in BIM: BIM workflows are increasingly or rapidly incorporating machine learning (ML) and artificial intelligence (AI). Large volumes or numerous data can be analyzed by AI algorithms, which can also be used to predict project final outcomes, identify patterns, trends, and anomalies, and facilitate collision detection. Machine learning (ML) algorithms can learn from past or historical data to increase or improve the precision of cost estimation, scheduling, and risk assessment.

7. Internet of Things (IoT) Integration: IoT and BIM function together to create "smart" infrastructure and buildings. Real-time data is captured by IoT sensors installed in assets, and this data can be connected to BIM models. Predictive maintenance, continuous monitoring, and better asset performance are made feasible by this integration.

8. Virtual and Augmented Reality (VR/AR): VR and AR technology improve collaboration and BIM visualization. In contrast to AR, which superimposes or overlays BIM models onto the real world, VR enables stakeholders to fully immerse themselves in virtual environment. These technologies provide maintenance services or activities, on-site visualization, clash detection, and design reviews.

9. Blockchain in BIM: Blockchain technology is being explored for safe, decentralized BIM data management. Blockchain can improve collaboration among project participants, the integrity of data, traceability, and transparency by minimizing disputes.

10. Digital Twin: Using real-time data from Internet of Things sensors, digital twin technology creates a virtual replica image of a physical asset or system. Digital twin creation and maintenance are made possible by BIM, which also enables performance tracking or monitoring, predictive analytics, and well-informed decision-making throughout the lifecycle of the asset.

The AEC sector is changing and reshaping because of these trends and advancements in BIM technology, which also improve project outcomes, efficiency, and collaboration. They are quite promising for BIM's future and its ability to help create built environments that are smarter and more sustainable.

13.2 BIM's role in the era of smart cities and digital twins

A smart city prioritizes efficiency, sustainability, and the welfare of its citizens by maximizing resource management, enhancing urban life, and improving public services through the use of digital technology and data-driven solutions. While a digital twin, on the other hand, is a virtual replica of a real-world system, object, or environment that is used to simulate, track, or monitor, and analyze real-world conditions in order optimize performance optimization and decision-making.

Building information modeling, or BIM, is essential in the era of digital twins and smart cities because it offers a thorough and integrated method of managing the built environment. The following is how BIM fits into these concepts.

1. Smart City Planning and Design: Smart city planning, and design are made possible by BIM. It enables the creation of virtual models of whole urban areas by architects and planners, incorporating a variety of data sources, including GIS, transportation networks, infrastructure, and environmental factors. BIM aids in planning for sustainable development, evaluating the effects of various design scenarios, and optimizing urban layouts.

2. Integrated Data and Systems: BIM acts as a central repository for digital information about assets, infrastructure, and assets in cities. BIM provides digital twins with the necessary data from several systems, along with IoT sensors. Digital twins are real-time, virtual replicas of physical systems or assets that can be monitored, analyzed, and simulated. BIM gives digital twins the information and context they need to support predictive maintenance, performance optimization, and well-informed decision-making.

3. Lifecycle Management: In a smart city, BIM supports an asset's whole lifecycle. Building Information Modeling (BIM) makes sure that accurate and up-to-date information is available at all stages of an asset's life, from the initial design and construction stages to ongoing operations, maintenance, and final renovation or decommissioning. By providing specifications, warranties, maintenance schedules, and other pertinent data, this information enables efficient lifecycle management and minimizes costs.

4. Real-Time Monitoring and Analytics: Real-time asset monitoring in a smart city is made possible by the combination of BIM, IoT, and sensor technologies. BIM models can incorporate data from sensors integrated into infrastructure, utilities, and buildings. Continuous monitoring of occupancy levels, energy usage, asset performance, and other parameters is made possible by this integration. After that, this data may be processed by analytics tools to find trends, forecast maintenance requirements, allocate resources optimally, and improve overall operational effectiveness.

5. Urban Simulation and Visualization: Smart city contexts benefit from the integration of BIM in urban simulation and visualization. Stakeholders can test the effects of changes, simulate alternative or different scenarios, and assess the effectiveness of different systems by building virtual models of the city. Combining BIM with cutting-edge visualization tools like augmented reality (AR) and virtual reality (VR) enables decision-makers to evaluate and experience planned developments, enhancing public engagement and communication.

6. Interoperability and Collaboration: The different stakeholders involved in the development and management of smart cities benefit from BIM's promotion of collaboration and interoperability. It facilitates coordination among contractors, facility managers, architects, engineers, urban planners, and local authorities by offering a standardized platform for information sharing. The capacity of BIM to incorporate data from various sources improves communication and facilitates effective decision-making.

In the end, BIM plays a crucial role in creating and managing intelligent urban environments in the modern era of digital twins and smart cities. Planning, design, and lifecycle management are supported, real-time analytics and monitoring are made possible, urban simulation, modeling, and visualization are made easier, and collaboration and interoperability are encouraged. BIM's capacity to promote sustainable and effective urban development is further enhanced by its integration with cutting-edge technologies.

13.3 Exploring the potential of artificial intelligence and machine learning in BIM

In building design and construction, artificial intelligence refers to the application or use of intelligent algorithms and systems that automate, optimize, and enhance building design, construction, and management or administration processes, hence improving decision-making and efficiency. On the other hand, machine learning in building information modeling (BIM) uses data-driven models to forecast outcomes, optimize designs, and identify patterns, allowing for continuous improvement in building performance and construction planning.

The application of Artificial Intelligence (AI) and Machine Learning (ML) has the potential to completely change how Building Information Modeling (BIM) is used. In this section, a number of important areas where AI and ML might improve BIM will be discussed in detail.

Through the analysis of BIM models, AI algorithms can detect conflicts between various building elements and automate the process of clash detection. Algorithms can be trained or taught using machine learning techniques or methods on previous clash data, allowing them to predict or forecast potential clashes and suggest solutions. This reduces errors, streamlines the clash resolution procedure, and enhances coordination in general (Clash Detection and Resolution).

Also, building designs can be optimized with the help of AI and ML. AI algorithms are able to generate multiple design options that satisfy specified criteria by analyzing large amounts of data and using generative design techniques. These algorithms enable engineers and architects to explore creative or innovative design ideas or solutions that optimize performance and reduce environmental effect by considering variables like cost, material usage, energy efficiency, and structural integrity (Design Optimization).

Energy Analysis and Simulation: AI and ML can improve energy analysis and simulation in BIM. ML algorithms can learn from historical energy usage data to create predictive models that estimate energy consumption in buildings. This enables designers to optimize energy efficiency, evaluate the impact of different design strategies, and make informed decisions regarding heating, cooling, lighting, and insulation systems.

In addition, Predictive maintenance of buildings and infrastructure can be made possible or using AI and ML. ML algorithms can identify or pinpoint patterns and anomalies that point or identify to potential or possible equipment breakdowns, failures, or maintenance needs by examining sensor data and previous maintenance records. This makes it possible for facility managers to schedule or plan maintenance activities in advance, allocate resources as efficiently as possible, and prevent costly unplanned downtime (Predictive Maintenance).

In construction projects, resource management and material utilization can be optimized with the help of AI and ML algorithms. Algorithms can recommend ideal quantities and efficient procurement methods or strategies by analyzing historical data on utilization of resources, waste generation, and material consumption. This reduces material waste, controls costs, and enhances sustainability (Material Optimization and Resource Management).

And cost estimation and project planning can be more accurate with the use of AI and ML. ML algorithms are able to analyze historical project data, such as schedules and cost records, to find patterns and generate more precise time and cost estimates for future projects. Better project planning, risk assessment, and budgeting are facilitated by this (Cost Estimation and Project Planning).

Construction projects can enhance safety and risk assessment with the use of AI and ML. To find potential safety hazards and forecast or predict risks, machine learning algorithms can analyze incident reports, project data, and historical safety records. This makes it possible to apply preventive actions or measures to enhance worker safety on construction sites and to adopt proactive risk management strategies (Safety and Risk Assessment).

While natural language processing methods driven by AI can allow chatbots and voice assistants to communicate or interact with BIM models. This gives stakeholders the ability to use voice commands or text-based communication to query the model, retrieve information, and do analysis. Natural language processing enhances BIM models' usability and accessibility for users across different disciplines. This is known as Natural Language Processing and Voice Assistants.

In a nutshell, there is great potential for enhancing several aspects of the construction lifecycle through the integration of AI and ML technologies with BIM. In the construction industry, AI and ML can enhance BIM workflows, improve decision-making, and increase efficiency and sustainability through a variety of applications, including clash detection, design optimization, energy analysis, predictive maintenance, and resource management.

In conclusion, the design, construction, and management of buildings or infrastructures have been completely transformed by BIM. This comprehensive book guide has provided beginners or novices to digital construction a solid foundation for understanding the principles or concepts and procedures of BIM. Professionals in the AEC sector can unlock new opportunities for innovation, increase efficiency, and improve collaboration by embracing BIM. Even if you're a student starting or just beginning your architectural or construction career or an experienced construction professional adjusting to the digital world or trend, the insights you acquire from this book guide will enable or equip you to fully utilize BIM and make a positive impact on the built environment going forward.

Appendix A
Glossary

3D BIM

Refers to the use of BIM technology to create a three-dimensional digital representation of a building's physical and functional characteristics. 3D BIM provides detailed geometry, spatial relationships, and geographic information, enabling better visualization and understanding of the design

4D BIM

An extension of 3D BIM that includes time-related information, such as scheduling and sequencing of construction activities. 4D BIM allows for the visualization of the construction process over time, improving project planning and coordination.

5D BIM

An extension of 4D BIM that includes cost-related information, enabling the estimation and tracking of project costs throughout the construction process.

6D BIM

Incorporates sustainability and energy analysis into BIM, focusing on the environmental impact, energy efficiency, and lifecycle management of a building.

7D BIM

An extension of 3D BIM that includes time-related information, such as scheduling and sequencing of construction activities. 4D BIM allows for the visualization of the construction process over time, improving project planning and coordination.

8D BIM

Focuses on health and safety management within the BIM framework. 8D BIM includes the identification and mitigation of potential hazards, improving safety planning and reducing risks during the construction process

9D BIM

Integrates information related to lean construction concepts or principles efficiency of construction production process. 9D BIM aids in proper management of resources, streamlining construction process by eliminating unnecessary activities and optimizing construction project outcomes by reducing waste and save cost.

10D BIM

Encompasses the use of BIM for industrialized construction, such as modular construction and prefabrication. 10D BIM facilitates the integration of manufacturing processes into the construction workflow, enhancing efficiency and reducing waste.

As-Built Model

A BIM model that represents the final, constructed state of a building, incorporating all changes and modifications made during construction. The As-Built Model serves as an accurate record of the completed project, useful for future renovations, maintenance, and facility management.

BIM Execution Plan (BEP)

A document that outlines the processes, roles, responsibilities, and standards for implementing BIM on a specific project.

Building Information Modeling (BIM)

A digital representation of the physical and functional characteristics of a building, used for planning, designing, constructing, and managing projects.

Clash Detection

A process in BIM where different models (e.g., architectural, structural, and MEP) are compared to identify and resolve potential conflicts or "clashes" before construction begins.

COBie (Construction Operations Building Information Exchange)

A standardized data format used in BIM to exchange building information, particularly for facilities management and operations.

Common Data Environment (CDE)

A centralized platform where all project information is stored, shared, and managed, ensuring that all stakeholders have access to up-to-date and accurate data.

Digital Twin

A digital replica of a physical building or infrastructure that is used for simulation, monitoring, and analysis throughout its lifecycle.

Federated Model

A composite model created by combining multiple individual BIM models (e.g., architectural, structural, MEP) to enable coordinated analysis and decision-making.

Geographic Information System (GIS)

A system that integrates spatial and geographical data into the BIM model, providing context for site analysis and planning.

Industry Foundation Classes (IFC)

An open file format and standard for exchanging BIM data across different software platforms, ensuring interoperability.

ISO 19650

An international standard for managing information over the lifecycle of a construction project using BIM.

Interoperability

The ability of different software systems and tools to work together and exchange data effectively within a BIM environment.

Laser Scanning

A technology used to capture detailed 3D data of physical spaces, often used in conjunction with BIM for creating accurate digital models of existing structures.

Level of Development/ Detail (LOD)

A measure of the amount of detail in a BIM model, ranging from LOD 100 (conceptual design) to LOD 500 (as-built conditions).

Lifecycle Assessment (LCA)

The evaluation of the environmental impacts associated with all stages of a building's life, from material extraction to demolition.

Model Authoring

The process of creating and developing a BIM model using specialized software tools, typically performed by architects, engineers, or contractors.

Parametric Modeling

A feature of BIM that allows for the creation of models where the relationships between elements are defined, enabling automatic updates when changes are made.

Point Cloud

A collection of data points in space, typically gathered through 3D laser scanning, used to create accurate 3D models of existing structures.

Quantity Takeoff

The process of extracting quantities of materials and components from a BIM model for estimating costs and planning procurement.

Revit

A popular BIM software developed by Autodesk, widely used for architectural design, structural engineering, and construction.

Scan to BIM

The process of capturing physical space data using 3D laser scanning and converting it into a digital BIM model.

Sustainability

In the context of BIM, refers to the incorporation of environmental and energy considerations into the design, construction, and operation of buildings.

Visualization

The use of 3D models to create visual representations of a building or structure, often used for presentations, simulations, and design validation.

Virtual Design and Construction (VDC)

The use of BIM and other digital tools to plan and manage construction projects in a virtual environment, improving coordination and reducing risks.

Virtual Reality (VR)

A technology that allows users to immerse themselves in a virtual representation of a BIM model, often used for design review and client presentations.

Work Breakdown Structure (WBS)

A hierarchical decomposition of a construction project into manageable sections, often used in conjunction with BIM for project planning and control.

Appendix B

List of Acronyms

AEC - *Architecture, Engineering, and Construction*
AECO - *Architects, Engineering, Construction, Operation*
AI- *Artificial Indigence*
API- *Application Programming Interface*
BEP - *BIM Execution Plan*
BIM - *Building Information Modeling*
BREEAM- *Building Research Establishment Assessment Methodology*
CAD - *Computer Aided Design*
CDE - *Common Data Environment*
COBie - *Construction-Operations Building Information Exchange*
CPI - *Cost Performance Index*
DCO – *Design, Construction, and Operation*
DMS - *Document Management System*
EIR - *Employer's Information Requirements*
GIS - *Geographic Information System*
IDM – *Information Delivery Manual*
IFC - *Industry Foundation Class*
IFD – *International Framework for Dictionaries*
IPD - *Integrated Project Delivery*
ISO - *International Organization for Standardization*
IT - *Information Technology*
LCA - *Lifecycle Assessment*
LEED- *Leadership in Energy and Environmental Design*
LOD - *Level of Development (Detail)*
MEP - *Mechanical, Electrical, and Plumbing*
ML- *Machine Learning*
NBS - *National Building Specification*
PIM - *Project Information Model*
PMI - *Project Management Institute*
O&M - *Operations and Maintenance*
QR - *Quick Response*
RFP - *Request for Proposal*
RFI - *Request for Information*
RFID - *Radio-Frequency Identification*
ROI - *Return on Investment*

VDC - *Virtual Design and Construction*
VR - *Virtual Reality*
WBS - *Work Breakdown Structure*
WIP - *Work in Progress*
3D - *Three-Dimensional (Spatial Coordination)*
4D - *Four-Dimensional (Time/ Schedule)*
5D - *Five-Dimensional (Cost/ Estimation)*
6D - *Six-Dimensional (Sustainability)*
7D- *Seven-Dimensional (Facility Management)*
8D- *Eight-Dimensional (Safety)*
9D- *Nine Dimensional (Lean Construction)*
10D- *Ten Dimensional (Industrial/Prefabrication)*

Appendix C

Resources for Further Learning

A. Books

1. BIM Handbook: A Guide to Building Information Modeling for Owners, Designers, Engineers, Contractors, and Facility Managers by Chuck Eastman, Paul Teicholz, Rafael Sacks, and Kathleen Liston.
2. Building Information Modeling: Planning and Managing Construction Projects with 4D CAD and Simulations by Willem Kymmell.
3. BIM and Construction Management: Proven Tools, Methods, and Workflows by Brad Hardin and Dave McCool.

B. Online Courses

1. Coursera - BIM Fundamentals for Engineers (offered by National Taiwan University).
2. edX - BIM for Smart Construction (offered by the Hong Kong University of Science and Technology).
3. Udemy - Revit Architecture: Beginner to Advanced (covers key BIM concepts and practices).

C. Professional Organizations

1. BuildingSMART International - Offers resources, standards, and training for BIM implementation globally.
2. National Institute of Building Sciences (NIBS) - Provides a wealth of information on BIM standards and best practices.
3. American Institute of Architects (AIA) - Offers guides and continuing education on BIM for architects.

D. Websites and Blogs

1. The BIM Hub - An online community and resource center for all things BIM.
2. AECbytes - Articles, case studies, and reviews focused on BIM and other AEC technologies.
3. BIM+ - News and insights on BIM developments across the construction industry.

E. Software Tutorials

1. Autodesk University - Free tutorials and classes on Autodesk software, including Revit, a leading BIM tool.
2. Graphisoft Learn - Training resources for ArchiCAD, a popular BIM software.
3. Bentley LEARNserve - Comprehensive training for Bentley's suite of BIM tools, including AECOsim Building Designer.

F. Certification Programs

1. RICS BIM Manager Certification - A globally recognized certification for BIM professionals offered by the Royal Institution of Chartered Surveyors.
2. Autodesk Certified Professional - Certifications for various Autodesk products, including Revit, that demonstrate BIM proficiency.
3. BuildingSMART Professional Certification - Certifies BIM skills in alignment with international standards.

G. Journals and Publications

1. Journal of Building Information Modeling (JBIM) - Covers the latest research and case studies in BIM.
2. Automation in Construction - A journal focusing on advanced construction technology, including BIM.
3. Engineering, Construction and Architectural Management - Research articles and case studies on BIM and other digital tools in construction.

Appendix D

Reference and Bibliography

1. **ABDELHAI, N.; 2022.** *Integration BIM and Emerging Technologies in Architectural Academic Programs. In book: Building Information Modeling - A Sustainable Approach and Emerging Technologies DOI:10.5772/intechopen.106443.*

2. **ALMASHJARY, B.M., ZOLKAFLI, U.K, RAZAK, A.S.A; 2020.** *Establishing Key Factors Towards Implementing Integrated Building Information Modeling (BIM) In AEC Industry: Malaysia Study. Journal of Building Performance, Vol: 11(1), ResearchGate publication, pp 117-126.*

3. **ANTONOPOULOU, S., BRYAN, P.; 2017.** *Historic England 2017 BIM for Heritage: Developing a Historic Building Information Model. Swindon. Historic England.*

4. **ARANDA-MENA, G., CRAWFORD, J., CHEVEZ, A., FROESE, T.; 2009**. *Building Information Modeling Demystified: Does It Make Business Sense to Adopt BIM? International Journal of Managing Projects in Business, 2009. Available at: [Google Scholar] (https://scholar.google.com/).*

5. **ARAYICI, Y., EGBU, C., COATES, P.; 2012**. *Building information modelling (BIM) implementation and remote construction projects: Issues, Challenges and Critiques, Journal of Information Technology in Construction (ITcon), Vol. 17, pg. 75 -92, http://www.itcon.org/2012/5*

6. **AUNG, T.H.S.; 2004**. *Dimensions in BIM: 7 BIM Dimensions Explained. NOVATR publication, (online) available at https://www.novatr.com/blog/dimensions-in-bim [Assessed 22/06/2023].*

7. **AZHAR, S.; 2011**. *Building Information Modeling (BIM): Trends, Benefits, Risks, and Challenges for the AEC Industry. Leadership and Management in Engineering, 2011. Available at: [Google Scholar] (https://scholar.google.com/).*

8. **AZHAR, S., HEIN, M., SKETO, B.; 2011.** *Building Information Modeling (BIM): Benefits, Risks, and Challenges. Journal of Construction*

Engineering and Management, 2012. Available at: [ResearchGate] (https://www.researchgate.net/).

9. **BARLISH, K., SULLIVAN, K.; 2012**. *How to Measure the Benefits of BIM — A Case Study Approach. Automation in Construction, 2012. Available at: [Internet Archive] (https://archive.org/).*

10. **BECERIK-GERBER, B., RICE, S.; 2010.** *The Perceived Value of Building Information Modeling in the U.S. Building Industry. Journal of Information Technology in Construction (ITcon), 2010. Available at: [ITcon](http://www.itcon.org/).*

11. **BOSCH-SIJTSEMA, P. M., GLUCH, P., SEZER, A. A.; 2019**. *Professional Development of the BIM Actor Role through Experiential Learning Automation in Construction, 2018. Available at: [Google Scholar] (https://scholar.google.com/).*

12. **BRISCOE, D.; 2016**. *Beyond BIM: Architecture Information Modeling. Routledge, New York, NY 10017.*

13. **BRYDE, D., BROQUETAS, M., VOLM, J. M.; 2012**. *The Project Benefits of Building Information Modeling (BIM). International Journal of Project Management, 2013. Available at: [Academia.edu] (https://www.academia.edu/).*

14. **CEROVSEK, T. A.; 2011**. *Review and Outlook for a 'Building Information Model' (BIM): A Multi-Standpoint Framework for Technological Development. Advanced Engineering Informatics, 2011. Available at: [Google Books] (https://books.google.com/).*

15. **CHIEN, K. F., WU, Z. H., HUANG, S. C.; 2014.** *Identifying and Assessing Critical Risk Factors for BIM Projects: Empirical Study. Automation in Construction, 2014. Available at: [ResearchGate](https://www.researchgate.net/).*

16. **CHONG, H. Y., LEE, C. Y., WANG, X.; 2017**. *A Mixed Review of the Adoption of Building Information Modeling (BIM) for Sustainability. Journal of Civil Engineering and Management, 2017. Available at: [Academia.edu] (https://www.academia.edu/).*

17. **DAVIES, R., HARTY, C.; 2013**. *Implementing 'Site BIM': A Case Study of ICT Innovation on a Large Hospital Project. Automation in Construction, 2013. Available at: [Google Scholar] (https://scholar.google.com/).*

18. **EASTMAN, C., TEICHOLZ, P., SACKS, R., & LISTON, K.; 2011**. *BIM Handbook: A Guide to Building Information Modeling for Owners, Designers, Engineers, Contractors, and Facility Managers*. John Wiley & Sons, 2011. Available at: [Open Library] (https://openlibrary.org/).*

19. **FISCHER, M., KUNZ, J.; 2004**. *The Scope and Role of Information Technology in Construction. Automation in Construction, 2004. Available at: [Google Scholar] (https://scholar.google.com/).*

20. **FORUM.AIPRM, AIPRM.COM.; 2023**. Help Forum and aid to writers (online) available at https://www.aiprm.com/.

21. **FUTURE LEARN; 2023.** *The four pillars of BIM. An Introduction to Building Information Modelling, Coventry University online affiliation course, (online) available at https://www.futurelearn.com/info/courses/building-information-modelling/0/steps/41384#:~:text=This%20article%20describes%20how%20the,%2C%20process%2C%20and%20technology). [Accessed July 15th, 2023].*

22. **GU, N., LONDON, K.; 2010**. *Understanding and Facilitating BIM Adoption in the AEC Industry. Automation in Construction, 2010. Available at: [ResearchGate] (https://www.researchgate.net/).*

23. **HAO, P.; 2019**. *Spatial Analysis. Hong Kong Baptist University, Hong Kong, The Wiley Blackwell Encyclopedia of Urban and Regional Studies. Edited by Anthony Orum. © 2019 John Wiley & Sons Ltd. Published 2019 by John Wiley & Sons Ltd; ResearchGate; DOI: 10.1002/9781118568446.eurs0306.*

24. **HARDIN, B., MCCOOL, D.; 2015.** *BIM and Construction Management: Proven Tools, Methods, and Workflows. Wiley, 2015. Available at: [Internet Archive] (https://archive.org/).*

25. **HUSSIEN, M. B.; 2023**. *Difference Between CAD and BIM. Revit Course, Faculty of Fine Arts-Architectural department, Scribd publication.*

(online) available at https://www.scribd.com/document/367206368/Difference-between-CAD-and-BIM-docx [Assessed July 14th, 2023].

26. **INGIBJORG et al.; 2017**. *Building Information modelling (BIM). Civil Engineering Faculty of Warsaw University of Technology, Warsaw 2017, Iceland, Great Britain.*

27. **KELLY, et al.; 2013**. *BIM for Facility Management: A Review and a Case Study Investigating the Value and Challenges. Proceedings of the 13th International Conference on Construction Applications of Virtual Reality, 2013. Available at: [ResearchGate](https://www.researchgate.net/).*

28. **KENSEK, K. M.; 2014**. *Building Information Modeling: BIM in Current and Future Practice. Wiley, 2014. Available at: [Internet Archive] (https://archive.org/).*

29. **KHAERUL, B. N.; 2023**. *BIM 360 Design Technical Manual 2. CAD PRO systems Ltd, and AUTODESK, Gold partner, Academia.edu publications (online) available at https://www.academia.edu/40565552/BIM_360_Design_Technical_Manual_2 [Assessed 07/08/2023].*

30. **KHOSROWSHAHI, F., ARAYICI, Y.; 2012**. *Roadmap for Implementation of BIM in the UK Construction Industry. Engineering, Construction and Architectural Management, 2012. Available at: [ResearchGate](https://www.researchgate.net/).*

31. **KLOTZ, L. E., HORMAN, M. J., & RILEY, D. R.; 2008**. *Process Transparency for Sustainable Building Delivery. Journal of Green Building, 2007. Available at: [Academia.edu] (https://www.academia.edu/).*

32. **KYMMELL, W.; 2008.** *Building Information Modeling: Planning and Managing Construction Projects with 4D CAD and Simulations. McGraw-Hill, 2008. Available at: [Google Books] (https://books.google.com/).*

33. **LIBRARY ADMIN; 2024**. *Project BIM Workflow. Creately, (online), available at https://creately.com/diagram/example/iv912pll1/project-bimworkflow#:~:text=A%20BIM%20workflow%20is%20a,all%20stages%20of%20a%20project. [Accessed January 9th, 2024].*

34. **LUCIANI, C. S., GARAGNANI, S., MINGUCCI, R; 2012**. *BIM TOOLS AND DESIGN INTENT: Limitations and Opportunities. Conference: Practical BIM 2012 - The USC BIM Symposium, ResearchGate publication, pp 1-27.*

35. **LUTH, G. P., SCHORER, A., TURKAN, Y.; 2013**. *Lessons from Using BIM to Increase Design-Construction Integration. Practice Periodical on Structural Design and Construction. 19. 10.1061/(ASCE)SC.1943-5576.0000200.*

36. **LINDEROTH, H. C. J.; 2010**. *Understanding Adoption and Use of BIM as the Creation of Actor Networks. Automation in Construction, 2010. Available at: [ResearchGate](https://www.researchgate.net/).*

37. **LIU, X., WANG, L., JI, G.; 2022**. *Optimization Approaches in Performance-Based Architectural Design A comparison study. 10.52842/conf.ecaade.2022.1.599.*

38. **LOVE, et al.; 2014.** A Benefits Realization Management Building Information Modeling Framework for Asset Owners. Automation in Construction. 37: pp. 1-10. *Available at: [https://espace.curtin.edu.au/handle/20.500.11937/16519]*

39. **MAHALINGAM, A., KASHYAP, R., & MAHAJAN, C.; 2010**. *An Evaluation of the Applicability of 4D CAD on Construction Projects. Automation in Construction, 19 (2), pp 148-159. ISSN 0926-5805; ELSVIER, Sciencedirect; https://doi.org/10.1016/j.autcon.2009.11.015*

40. **MCGRAW HILL CONSTRUCTION.; 2014**. *The Business Value of BIM for Construction in Major Global Markets: How Contractors Around the World Are Driving Innovation With Building Information Modeling. McGraw-Hill, 2014. Available at: [Internet Archive] (https://archive.org/).*

41. **MCPARTLAND, R.; 2017**. *What are BIM objects? NBS publication, (online) available at https://www.thenbs.com/knowledge/what-are-bim-objects [Accessed July 14th, 2023].*

42. **MIETTINEN, R., PAAVOLA, S.; 2014**. *Beyond the BIM Utopia: Approaches to the Development and Implementation of Building Information Modeling. Automation in Construction, 2014. Available at: https://www.academia.edu/7273375/Beyond_the_BIM_utopia_Approaches_to_the_development_and_implementation_of_building_information_modeling_2014_.*

43. **MELOVICS, C.; 2022**. *What are the BIM dimensions? BUILDEXT publication, (online) available at https://buildext.com/en/bim-dimensions/ [Assessed 22/06/2023].*

44. **MIRAJKAR, S.; 2021**. *Significance of Planning Tools in BIM. Architect and BIM professional, Linkedin publication, (online) available at https://www.linkedin.com/pulse/significance-planning-tools-bim-shraddha-mirajkar [Assessed 08/08/2023].*

45. **MORENO et al.; 2022**. *Dynamic Data Feeding into BIM for Facility Management: A Prototype Application to a University Building. Buildings; Vol: 12(5):645. https://doi.org/10.3390/buildings12050645*

46. **NGUYEN, T., NGUYEN, T., TRAN, T.; 2024.** *Building Information Modeling (BIM) for Construction Project Schedule Management: A Review. Engineering, Technology & Applied Science Research. 14. 13133-13142. 10.48084/etasr.6834.*

47. **POIRIER, E. A., STAUB-FRENCH, S., FORGUES, D.; 2015**. *Measuring the Impact of BIM on Labor Productivity in a Small Specialty Contracting Enterprise through Action-Research. Automation in Construction, 2015. Available at: https://www.academia.edu/14882060/Measuring_the_impact_of_BIM_on_labor_productivity_in_a_small_specialty_contracting_enterprise_through_action_research.*

48. **SACKS, et al.; 2010**. *The Rosewood experiment — Building information modeling and interoperability for architectural precast facades, Automation in Construction, Vol: 19(4), Pp 419-432, ISSN 0926-5805, ELSEVIER, ScienceDirect; https://doi.org/10.1016/j.autcon.2009.11.012.*

49. **SANTOS, E.; 2010**. *Building Information Modeling and Interoperability. SIGraDi 2009 - Proceedings of the 13th Congress of the Iberoamerican Society of Digital Graphics, Sao Paulo, Brazil, November 16-18, 2009.*

50. **SEBASTIAN, R., BERLO, L. V.; 2010**. *Tool for Benchmarking BIM Performance of Design, Engineering and Construction Firms in The Netherlands. Architectural Engineering and Design Management, 2010. Available at: https://www.academia.edu/1477483/Tool_for_Benchmarking_BIM_Performance_of_Design_Engineering_and_Construction_Firms_in_The_Netherlands.*

51. **SHEN et al.; 2009**. *Systems Integration and Collaboration in Architecture, Engineering, Construction, and Facilities Management: A Review. Automation in Construction, 2010. Available at: https://www.researchgate.net/publication/220371481_Systems_integration_and_collaboration_in_architecture_engineering_construction_and_facilities_management_A_review.*

52. **SHOU et al.; 2015**. *Comparative Review of Building Information Modelling Implementation in Building and Infrastructure Industries. Archives of Computational Methods in Engineering. Vol: 22: pp. 291-308. Available at: https://espace.curtin.edu.au/handle/20.500.11937/35414.*

53. **SMITH, D. K. TARDIF, M.; 2009.** *Building Information Modeling: A Strategic Implementation Guide for Architects, Engineers, Constructors, and Real Estate Asset Managers. Wiley, 2009. Available at: https://books.google.com.ng/books/about/Building_Information_Modeling.html?id=vr2Vccw-0soC&redir_esc=y.*

54. **SOLIBRI; 2016.** *Imagine: Reliable Information on Demand. About Solibri on www.solibri.com, Solibri UK Ltd publication, (online) available at https://www.pbctoday.co.uk/news/wp-content/uploads/2015/05/Solibri-ebook-6141-small.pdf [Assessed 07/08/2023].*

55. **SUCCAR, B.; 2016**. *BIM Framework Essentials - BIM Maturity. presentation (and video) introduces BIM Maturity, the gradual and continual improvement in quality, repeatability and predictability within a BIM Capability, Academia.edu publications, (online) available at*

https://www.academia.edu/16553750/BIM_Framework_Essentials_BIM_Maturity [Assessed 22/06/2023].

56. **SUCCAR, B.; 2010.** *Building Information Modelling Maturity Matrix. In book: Handbook of Research on Building Information Modelling and Construction Informatics: Concepts and Technologies, ResearchGate publication. 10.4018/978-1-60566-928-1.ch004.*

57. **SUCCAR, B.; 2010**. *The Five Components of BIM Performance Measurement. Conference: CIB World Congress, ResearchGate publication, 10.13140/2.1.3357.1521.*

58. **SUCCAR, B.; 2009**. *Building Information Modeling Framework: A Research and Delivery Foundation for Industry Stakeholders. Automation in Construction, 2009. Available at: [ResearchGate] (https://www.researchgate.net/).*

59. **SUGENG et al.; 2024.** *Integration of Building Information Modelling (BIM) in Civil Engineering Project: A Literature Review. Indonesia Journal of Engineering and Education Technology (IJEET). Vol: 2; pp 319-328. 10.61991/ijeet.v2i2.56.*

60. **THE HOURIGAN TEAM; 2013**. *5 Fundamentals of BIM for Improved Design and Construction. Hourigan publication, (online) available at https://www.hourigan.group/blog/5-fundamentals-of-bim-for-improved-design-and-construction/ [Assessed 01/07/2023].*

61. **UNDERWOOD, J., ISIKDAG, U.; 2010**. *Handbook of Research on Building Information Modeling and Construction Informatics: Concepts and Technologies. IGI Global, 2010. Available at: https://www.academia.edu/208203/Handbook_of_Research_on_Building_Information_Modeling_and_Construction_Informatics_Concepts_and_Technologies.*

62. **VERMA, S.; 2022**. *Dynamo For Revit: Everything You Need to Know in (2023). NOVATR previously oneistox, Novatr Network Pvt. Ltd. Publication, (online) available at https://www.novatr.com/blog/dynamo-for-revit [Assessed 08/08/2023]*

63. **VOLK, R., STENGEL, J., SCHULTMANN, F.; 2014**. *Building Information Modeling (BIM) for Existing Buildings — Literature Review and Future Needs. Automation in Construction, 2014. Available at: [ResearchGate](https://www.researchgate.net/).*

64. **WU, W., ISSA, R. R. A.; 2012**. *BIM-Enabled Building Commissioning and Handover. Journal of Building Information Modeling (JBIM), 2015. Available at: https://www.researchgate.net/publication/264159184_BIM-Enabled_Building_Commissioning_and_Handover.*

Appendix E

Author Biography

Goodluck Eromonsele is a seasoned professional in the field of construction and facilities management, with a strong academic background and extensive practical experience. He holds a Master of Science (MSc) in Construction Project Management from The Scott Sutherland School of Architecture and Built Environment, Robert Gordon University, United Kingdom, where he developed expertise in various aspects of construction, including Building Information Management (BIM), sustainability, construction project cost estimation, and design management. His academic journey began with an undergraduate degree in Building Construction Technology, which he completed with distinction.

In addition to his formal education, Goodluck has expanded his knowledge in Building Information Modelling (BIM) through specialized training. He completed the Alison BIM Academic training program, earning a certificate (AC-5429-32082286) in "Basics of Building Information Modelling (BIM)." His coursework included modules on the introduction to BIM and its application in design, construction, and ownership.

Goodluck has also contributed to the academic community with several research articles published online. Notable works include *Building Information Modelling for Construction Project Managers: A Bibliometric Analysis Approach*. His research focuses on the integration of BIM in construction management, with a particular interest in the challenges and opportunities within the construction sector.

With his deep knowledge and passion for the construction industry, Goodluck Eromonsele continues to influence and contribute to the advancement of construction project management practices, particularly in the adoption and implementation of BIM.

www.ingramcontent.com/pod-product-compliance
Lightning Source LLC
LaVergne TN
LVHW022103200926
848481LV00003B/1052
* 9 7 9 8 8 9 9 6 6 4 2 6 7 *